UNLOCK

A JOURNEY THROUGH THE HALLMARKS OF GROWING OLD

THE BIOLOGY OF AGING AND THE NEW SCIENCE OF LONGEVITY

THE F(x) LONGEVITY SERIES
BOOK 0

F(x) Insights Publishing

Andrew Kain

TABLE OF CONTENTS

ABOUT THIS BOOK

U*NLOCK — A journey through the hallmarks of growing old. The biology of aging and the new science of longevity.*

The F(x) Longevity Series — Decode aging. Reclaim function. Book 0.

ABOUT THIS EDITION

This is the integrated Reader-Tool Edition of UNLOCK.

What the boxes do

CLOSER LOOK boxes go deeper than the narrative — a peer-reviewed mechanism, a landmark experiment, a careful caveat. Each carries an APA citation. EASY ACCESS boxes do the opposite: they translate. One short paragraph in plain language for anyone who wants the takeaway without the technical scaffolding. KEY TERMS boxes close each chapter with a working glossary of every term the chapter assumed.

How to read them

Boxes are optional. The narrative reads cleanly without them — they sit in the gutter of the page rather than the spine. Read straight through if you want the story; pause at the boxes if you want the receipts.

The three box styles

CLOSER LOOK *Sample — Closer Look*

Closer Look boxes deepen a claim, name the experiment behind it, and cite the source. Look for them after a passage where the science deserved more room than the narrative could give it.

Author, A. (Year). Title of the work. Journal, vol(issue), pages. https://doi.org/...

EASY ACCESS *Sample — Easy Access*

Easy Access boxes are the plain-language version of an idea. If a paragraph leans technical, the Easy Access box right after it says the same thing in everyday words.

KEY TERMS *Sample — Key Terms*

Term — A concise working definition — built for a reader who has not seen this word before but will see it again three chapters later.

Another term — Same idea: short, accurate, useful in context.

INTRODUCTION

THE BULLET HOLES WE DIDN'T SEE

In 1943, Allied warplanes were returning from bombing missions over Europe, riddled with bullet holes. Engineers gathered data on the damage, mapping out the most frequently hit sections of the aircraft—wings, fuselage, tail. The plan was obvious: reinforce those areas with extra armor. But a Hungarian-born mathematician named Abraham Wald disagreed.

"You're looking at the planes that made it back," Wald said. "You should be reinforcing the areas with no bullet holes. Because those are the places where, if hit, the plane didn't survive."

It was a moment of inverted logic so elegant it seems obvious in hindsight. The damage wasn't the message—it was the silence that mattered. The pattern not seen was the one that told the truth.

We begin here not because this is a book about aircraft or war. But because aging science, for much of its history, made the same kind of mistake. We focused on what we could see—the diseases that show up in late life: heart attacks, strokes, cancer, Alzheimer's. We charted their causes, catalogued their symptoms, and raced to develop cures. We mapped the bullet holes on the planes that made it home.

But we didn't ask the deeper question: What about the system that didn't make it back at all?

Only recently have scientists begun to look at what underlies those visible failures. Not the symptoms, but the substrate. Not the tumors or plaques or weakened bones—but the molecular processes that make all of them more likely with each passing year. Inflammation that never turns off. Mitochondria sputtering like old engines. Cells that retire but don't leave. A web of interconnected breakdowns that, when left unchecked, becomes the scaffolding for every disease we call "age-related."

This book is about those invisible vulnerabilities. The bullet holes we didn't see. The systemwide collapses that accumulate slowly, silently—until one day, a heart attack or memory lapse becomes our version of a downed aircraft.

Aging, in this view, is not the inevitable backdrop of disease—it is the architecture of risk. And if we understand it, we may be able to reinforce the right parts.

WHAT WE THINK AGING IS — AND WHY THAT'S WRONG

Ask someone what aging is, and you'll likely get a folk answer. Graying hair. Aching knees. A slower gait. Maybe a few jokes about forgetfulness or stubborn grandfathers. The story is visual, superficial, and deeply ingrained. This is what aging looks like, we think. The body changes. The mind slows. It's natural. Unavoidable. Unremarkable.

Scratch a little deeper and you find the medical model—more technical, but no less linear. According to this model, aging is simply time's effect on tissue. The longer we live, the more our cells and organs deteriorate. That decline leads to disease, and eventually, death. End of story.

But both views share the same invisible premise: that aging is something that just happens. A universal, inertial drift toward dysfunction. In this narrative, the body is a machine wearing out, like a car racking up miles on the odometer. We can tune it up. Replace parts. But the fundamental trajectory never changes.

This is the fatalistic model. And it's wrong.

Not just wrong in theory—but wrong in the way Abraham Wald's engineers were wrong. They mistook what was visible for what was important. They looked at damage and thought they understood vulnerability. In the same way, our culture looks at wrinkles and memory loss and thinks it understands aging.

But science, especially over the last two decades, tells a more curious story. One that suggests aging isn't a slow unraveling of fate—but a dynamic process. A program with levers and switches. A set of biological systems that can accelerate, decelerate, and even—under the right conditions—be rewound.

This might sound fantastical. But so would insulin therapy to a 19th-century doctor watching diabetics waste away. So would immunotherapy to someone who once believed cancer was purely a death sentence. So would bone marrow transplants, in vitro fertilization, or replacing a diseased heart with a donor's.

We are at the edge of a similar shift in how we understand aging.

And to fully grasp it, we need to make a psychological shift first.

KEY TERMS *The vocabulary of the framework*

Hallmark of aging — A cellular or molecular process whose decline is sufficient to drive aging and whose restoration is sufficient to slow it. Twelve are currently recognized.

Healthspan — The years lived in good function, distinct from lifespan, which is years lived in any state.

Compression of morbidity — James Fries's 1980 hypothesis that medicine's goal should be to push the onset of chronic disease closer to death — narrowing the bad years rather than extending the average ones.

Geroscience — The branch of aging research that targets the biology of aging itself, rather than treating each age-related disease one at a time.

In his seminal book Thinking, Fast and Slow, Daniel Kahneman describes two modes of human cognition.

System 1 is fast, intuitive, emotionally charged. It jumps to conclusions. System 2 is slower, analytical, effortful. It questions.

When it comes to aging, we live almost entirely in System 1. We rely on metaphors—"the body as a candle burning down," "youth slipping away," "the circle of life." These feel true. They have moral and emotional weight. But they don't explain the biology.

System 1 also loves binaries: young vs. old, healthy vs. sick, alive vs. dying. But the biology of aging resists such neat categories. A 50-year-old with high-functioning mitochondria and robust proteostasis might biologically resemble a 35-year-old. A 40-year-old with chronic inflammation and rapid telomere loss might biologically resemble someone twice their age. The number on the birthday cake doesn't tell the story anymore.

System 2, however, invites a more nuanced perspective. It asks: *What if aging isn't a singular thing? What if it's a network of processes that can be observed, measured, and influenced?*

That's what this book is about. A deep exploration into the 12 hallmarks of aging—scientifically recognized mechanisms that explain how and why we age. Each chapter dives into one hallmark: DNA instability, mitochondrial dysfunction, telomere attrition, and so on. But none of them exist in isolation. They talk to each other. They reinforce each other. And in that interplay, the real story of aging begins.

So this is not a story of decay. It is a story of miscommunication. Of cellular confusion. Of systems losing coordination over time.

And—most radically—it's a story that can change.

Because if aging isn't a pre-written script but a dynamic conversation between cells, organs, and systems... then perhaps it can be edited. Not to erase our mortality, but to compress our suffering. Not to chase immortality, but to extend vitality.

We've reinforced the wrong bullet holes for too long. Now, we start asking:

What brought the whole plane down in the first place?

THE BIRTH OF A. NEW FRAMEWORK

In the spring of 1961, Leonard Hayflick was working with fetal human cells at the Wistar Institute in Philadelphia. The job was, frankly, tedious—grow cells in culture, monitor their replication, repeat. At the time, the prevailing belief was that human cells, once immortalized in a dish, could divide forever. It was a tidy theory with one problem: it wasn't true.

Hayflick noticed that after about 50 divisions, the cells simply stopped dividing. They didn't die. They didn't mutate. They just... froze. Alive but inert. He repeated the experiment. Same result. The idea that human cells had a built-in replication limit was heretical at the time. His colleagues dismissed it. Some scoffed. But Hayflick persisted.

That anomaly—dismissed as an error in cell handling—would eventually give birth to the concept of cellular senescence, now recognized as one of the most important hallmarks of aging. The phenomenon that bored his colleagues to tears would later become the basis for multi-billion-dollar pharmaceutical efforts to clear senescent cells and extend healthy lifespan in mice, dogs, and potentially humans.

This is how revolutions in biology begin: not with thunderclaps, but with quiet deviations. A culture dish. An unexpected stall. A few stubborn researchers who ask, What if this isn't noise? What if it's signal?

Fast-forward to 2013.

A landmark paper, The Hallmarks of Aging, is published by Carlos López-Otín and colleagues in the journal Cell. It does not announce a cure. It does not offer a drug. But it does something far more disruptive: it organizes the chaos. It proposes that aging isn't a fog of decline, but a system of identifiable and targetable processes—twelve, to be exact.

CLOSER LOOK ***What the hallmarks framework actually claims***

The original Hallmarks of Aging paper proposed nine cellular and molecular processes that, taken together, explain organismal aging. Each had to satisfy three criteria: it appears with normal aging; experimentally accelerating it shortens lifespan or healthspan; and experimentally attenuating it extends them. Genomic instability, telomere attrition, epigenetic alterations, loss of proteostasis, deregulated nutrient sensing, mitochondrial dysfunction, cellular senescence, stem cell exhaustion, and altered intercellular communication made the original list.

A 2023 update added three more — disabled macroautophagy, chronic inflammation, and microbiome disturbances — and emphasized that the hallmarks are not independent. They form a network: damage to one node propagates to others, and successful interventions tend to hit several at once. Aging, in this view, is what an organism looks like when its cellular control circuits start drifting out of phase with each other.

López-Otín et al. (2023). Hallmarks of aging: An expanding universe. Cell, 186(2), 243–278.

These included:

Genomic instability - Telomere attrition - Epigenetic alterations - Loss of proteostasis - Deregulated nutrient sensing - Mitochondrial dysfunction - Cellular senescence - Stem cell exhaustion - Altered intercellular communication - And later additions: chronic inflammation, disabled macroautophagy, and microbiome disturbances.

EASY ACCESS ***Why a list of twelve matters***

Before 2013, the science of aging was a fog of overlapping theories — wear and tear, oxidative damage, hormonal decline, programmed obsolescence. Each had a piece of the truth and none had the whole. The hallmarks framework did something simple but powerful: it named twelve discrete biological processes, each measurable, each potentially treatable, each shown to drive aging on its own. It turned aging from a mood into a checklist. That single move — inventory before intervention — is what made the next decade of geroscience possible.

Each process alone can age a tissue. But together, they form a biological web of interdependence. One breaks, another follows. A telomere shortens, a cell senesces, SASP factors inflame neighboring cells, stem cells go dormant, and tissue degenerates.

This framework did for aging what the periodic table did for chemistry: it offered structure. For the first time, researchers could map interventions, not to vague goals like \"anti-aging,\" but to precise cellular defects. The field shifted from delay to repair, from longevity as an abstract wish to aging as a biological puzzle—and potentially, a solvable one.

It also reframed aging as a disease of control. Not in the metaphorical sense, but in the systems biology sense. DNA repair systems fall behind. Protein quality control collapses. Immune signaling, once sharp and swift, becomes slow and sloppy. The cell doesn't fail because one part breaks—it fails because feedback fails. Aging, like cancer, is what happens when a once self-regulating system loses its ability to adapt, to rebound, to remember its own rules.

That's the paradox. Aging is not chaos. It's order—too much order. Senescent cells refuse to die. Proteins cling to their misshapen folds. Inflammatory circuits loop endlessly. What we call decline is often rigidity masquerading as entropy.

The 2013 framework gave the field its lexicon. But the real revolution was conceptual. It asked researchers—and, increasingly, the public—to see aging not as an immutable march toward death, but as a modular system that can be measured, understood, and potentially modified.

WHY THIS BOOK TAKES A NARRATIVE APPROACH

I didn't want to write another textbook.

There are already plenty—thick with molecular pathways, rodent lifespan curves, and graphs annotated in font sizes only a postdoc could love. They're important. But they aren't memorable. Not because the information isn't valuable—but because humans don't encode information in lists. We encode it in stories.

Think about what you remember from school. It wasn't the Krebs cycle. It was probably the story your teacher told about a scientist getting struck by inspiration—or lightning—or rejection letters. That's what sticks. Our memory was built for myths, not mechanisms.

That's why each chapter in this book begins with a narrative anchor. Sometimes it's a character: a marathoner in his nineties, a chef losing the dexterity of her hands, a patient entering a senolytic trial. Other times it's an evolutionary puzzle, or a piece of scientific history, or a parable drawn from a seemingly unrelated field—WWII statistics, ecological fire cycles, a symphony falling out of tune.

But they all serve the same purpose: to make complexity feel familiar.

Because aging is not one story—it's many. It's the story of our cells forgetting how to communicate. It's the story of the mitochondria powering down. It's the story of repair systems falling behind. And when we reduce it to a single metaphor—"wear and tear," "a ticking clock"—we lose the richness of what's actually happening beneath our skin.

So this book doesn't aim to simplify aging. It aims to humanize it.

To write it not in bullet points, but in full sentences. Not as inevitability, but as narrative.

Humans don't respond to numbers—they respond to frames. A risk framed as survival sounds different than the same risk framed as death. A fact framed as hope is processed differently than one framed as fear. Our minds are not spreadsheets. They're story engines.

That's why this book doesn't just present data. It frames it. It asks: How would you feel if a single pill could clear out your body's toxic cells? What would it mean to know that your immune system, not your joints, is the root cause of your fatigue? How do we rethink aging if it's not the enemy, but the map?

Aging, when reframed, becomes not a cliff—but a mountain. Steep, yes. But climbable. Navigable. And increasingly, re-routable.

That's the power of framing. It lets us see old problems in new ways. And often, that's all science ever really is: better metaphors with better measurements.

WHAT YOU'LL LEARN—AND WHAT YOU WON'T

Let's be clear from the outset: this book won't tell you how to live to 150. It won't promise immortality in a capsule or eternal youth via superfoods. You won't find a grocery list in Chapter 6, or a supplement stack backed by secret Nobel laureates. If that's what you're looking for, you've wandered into the wrong narrative.

What you will find is something much rarer: clarity. You'll learn what aging is—not just what it looks like.

You'll see how each process unfolds, where it begins, where it breaks down, and what we currently understand about how it might be slowed, buffered, or even reversed. But more than that, you'll see why these processes matter.

You'll learn that aging isn't a unified phenomenon—it's a collection of breakdowns. A mosaic of malfunctions. A system slowly losing its capacity for balance. One hallmark doesn't kill you. But the interaction between many—genomic instability triggering senescence triggering inflammation—creates a cascade that does.

You'll discover that aging is personal, cellular, and deeply entangled. You'll see how a single change—like improved autophagy—can ripple across mitochondrial health, immune surveillance, and even mood regulation. You'll come to understand that this is not theory anymore. It's happening. In labs. In trials. And in some cases, quietly, inside people walking among us.

You'll also see that we're still early. Many of the interventions discussed here—partial epigenetic reprogramming, senolytic drugs, NAD+ boosters, telomerase gene therapies—are nascent, often tested only in mice or specific diseases. Some will fail. Others will evolve. A few may quietly change the way we think about health and lifespan altogether.

And yes, you'll meet the critics too.

The researchers who argue that targeting aging might yield diminishing returns. That we may increase lifespan, but not healthspan. That we risk creating a world where the rich live to 110 and the poor die at 65. These are not fringe voices. They are necessary counterweights. Their skepticism keeps the science honest.

You will not find certainty here. But you will find orientation—a way to think about aging that replaces fear with understanding, despair with inquiry. That's all science can ever really promise: not rescue, but perspective.

Let me leave you with a scene.

It's thirty years from now. A quiet morning in late autumn. A woman in her early nineties opens her window, pours herself coffee, and heads to the park. Her knees don't ache. Her grip is strong. She walks briskly. Her gait is not the shuffle we've come to associate with the old, but the stride of someone still fully inhabiting her body.

In the park, she meets her granddaughter. They walk, talk, laugh. The granddaughter is studying cell biology—drawn in by a generation that, decades before, decided aging was no longer offlimits to science.

Later that day, the woman teaches a cooking class. Still works parttime. Still mentors. Still dreams.

She is not immortal. She is not untouched by time. Her hair is silver, her skin is lined. But her function—her vitality—is intact. Aging, in her case, has become what it always should have been: a gradual winding down, not a catastrophic collapse.

This future is not fantasy. Every intervention discussed in this book—every trial, every theory—has this goal: not to stop aging, but to reshape its curve. To delay dysfunction. To compress morbidity. To ensure that the final chapters of life are full of autonomy, contribution, and joy.

The science is still catching up. The debates are still fierce. But the direction is clear.

For the first time in human history, we are treating aging not as an inevitability—but as a design flaw.

And we are learning to edit.

Reader-tool boxes for this chapter

CLOSER LOOK ***Hayflick's heretical experiment***

In 1961 Leonard Hayflick and Paul Moorhead published a paper that quietly upended a generation of cell biology. They showed that normal human fibroblasts grown in culture divide a finite number of times — about fifty — and then enter a stable, non-dividing state. The dogma of the day, established by Alexis Carrel decades earlier, held that animal cells were essentially immortal in culture if you fed them well enough. Hayflick's result was treated with skepticism for years; it implied a built-in cellular limit nobody had asked for.

What he had actually found was the cellular face of aging itself. The phenomenon he observed would later be called replicative senescence, and the number became known as the Hayflick limit. Decades later, telomere shortening would supply the mechanism. But the conceptual leap was Hayflick's: the idea that aging is partly a property of cells, not just of organisms.

Hayflick & Moorhead (1961). The serial cultivation of human diploid cell strains. Experimental Cell Research, 25(3), 585–621.

THE BROKEN BLUEPRINT

GENOMIC INSTABILITY

Dr. Leena Patel wasn't supposed to notice anything unusual. She was a junior doctor, newly assigned to the oncology ward in a midsized hospital on the outskirts of Kyiv. The year was 2008. Her patients, mostly elderly, were being treated for standard fare—colon cancer, lymphoma, occasional cases of leukemia. But there was a subset of patients that didn't quite fit the pattern. They were younger than most. In their fifties or early sixties. Yet their bodies were crumbling in ways she associated with people much older: brittle skin, graying hair, teeth lost prematurely, cataracts that seemed far too aggressive for their age. Some had sarcopenic limbs, frail despite adequate nutrition. One man, a former construction worker, had lost two inches in height from unexplained vertebral compression fractures. Another, a woman who had never smoked, had developed a form of lung cancer usually found in patients twice her age.

Leena didn't know what to make of it. She asked questions. Had they worked with chemicals? Lived near factories? Eventually, a pattern began to form—not in occupation or behavior, but geography. They had all grown up near the Exclusion Zone. Not close enough to have been poisoned outright, but within a band of villages where fallout had drifted and settled after the 1986 Chernobyl explosion. For the children of that disaster, the effects had been sudden and visible. But for these adults, the damage seemed subtler, almost spectral. It wasn't just that they got sick. It was how they aged—fast, erratically, and irreversibly.

Over the next three years, Leena started documenting. She reviewed medical records, reached out to other hospitals, and pieced together a private case series. What emerged wasn't conclusive, but it was consistent: people exposed to chronic, low-dose radiation in their early years were experiencing an accelerated breakdown of their bodies. They weren't dying in large numbers. They were aging too fast. And the deeper she looked, the more it seemed that their problems shared a common origin—an invisible disintegration happening inside each cell, quietly corrupting the body's most basic instructions.

That was how she stumbled into the world of genomic instability.

EASY ACCESS ***What just happened, in plain language***

Genomic instability is the slow loss of fidelity in your DNA. Every day, the molecules that carry your body's instructions take small hits — from sunlight, from oxygen, from the simple act of copying themselves. Most of those hits are repaired. Some aren't. Over decades, the unrepaired errors pile up. Leena's Chernobyl-zone patients aged quickly because their cells took more hits than usual, and the repair crews could not keep pace. The rest of us live the slower version of the same story.

In medical school, Leena had been taught that DNA was a kind of biological manuscript—dense, precise, and remarkably self-correcting. She pictured it as a code etched into every cell, duplicated perfectly through trillions of cycles of replication. That metaphor, elegant as it was, had never fully captured the chaos of reality. DNA, she now saw, was less like a book and more like a blueprint kept in a machine shop, exposed to constant wear, smoke, and metal dust. It frayed. It bent. Pieces went missing.

And yet, somehow, life carried on. But only up to a point.

Leena began corresponding with researchers working on the long-term effects of radiation exposure. She read papers on Hiroshima survivors, on astronauts exposed to cosmic rays, on patients who had undergone early radiotherapy. In each case, the results hinted at a biological truth bigger than any specific incident: the genome, the very instruction set that builds and maintains a human body, does not remain pristine. It accumulates wounds—some shallow and repaired, others deep and miscopied. With time, those wounds begin to pile up. The cell, overworked and imprecise, starts to misfire.

This gradual accumulation of DNA damage, scattered across the genome, is what scientists now call genomic instability—the slow unraveling of the fidelity of our genetic code.

It happens in everyone, even those untouched by nuclear fallout. Every second, somewhere in your body, a strand of DNA is broken. Ultraviolet light from the sun nicks it. Replication machinery slips. Oxygen molecules, in their reactive form, bash into its delicate bonds. In total, your body sees tens of thousands of DNA lesions per day. Most are caught. Enzymes swarm the damage like emergency responders—snipping, patching, and sealing. But the system isn't perfect. As we age, the repair process becomes slower and more errorprone. Mistakes sneak through. Some of those mistakes matter. A single miswritten line in the genome of a stem cell can multiply into a thousand defective cells, each carrying a distorted copy of the original script.

The problem isn't just damage. It's the pattern of damage over time. The genome isn't a passive storage device. It's an active, interactive code—directing which proteins to make, when to divide, how to adapt. When that code is corrupted, the effects ripple outward. Cells become sluggish. Some lose their identity. Others grow recklessly, ignoring internal controls. This is how cancer begins. But not just cancer.

This is also how tissues age.

KEY TERMS *The vocabulary of damage*

Genome — The full instruction set carried in your DNA. About two meters of code, coiled into 46 chromosomes, present in nearly every cell.

Somatic mutation — A change in DNA that occurs after birth, in a single cell or its descendants. Not inherited from your parents and not passed to your children.

DNA lesion — A specific instance of damage — a break, a chemical alteration, a missing base. Most are repaired within minutes.

Stem cell — A long-lived cell that renews tissue by dividing into specialized daughter cells. Mutations here scale: one bad stem cell can seed thousands of bad descendants.

Tumor suppressor gene — A brake on cell division. Disable it, and a damaged cell may keep dividing when it should have stopped.

Oncogene — An accelerator of cell division. Activate it inappropriately, and a cell starts pressing the gas pedal it shouldn't.

What struck Leena most wasn't just the correlation between DNA damage and disease. It was the timing. The same mutations that caused leukemia in her patients were also appearing in their skin, bones, and muscle—long before the cancer took hold. These weren't isolated events. They were symptoms of a broader biological decay. Genomic instability wasn't merely about oncogenes and tumor suppressors. It was about entropy, silently eating away at the structural coherence of the body.

She found herself wondering: If the genome is a blueprint, what happens when enough of it becomes unreadable? Can a house still stand if a few bricks are missing? What about entire walls?

That led her to the studies of progeroid syndromes—genetic disorders like Werner syndrome and Hutchinson-Gilford progeria, where patients inherit mutations that prevent their DNA from being properly maintained. These children age at ten times the normal rate. By age thirteen, they often have the arteries of an octogenarian. The evidence is not subtle. When the genome can't be kept stable, the body cannot stay young.

But then another question emerged. If breakdown of the genome is a driver of aging, why do some people with high mutation loads seem to age normally? There were cases of healthy centenarians with surprisingly high somatic mutation burdens in their blood or colon tissue. And there were mice that, despite carrying thousands of small DNA mutations, remained robust—until something else, some other failure, tipped the balance.

The more Leena learned, the more she saw the limitations of a simple narrative. DNA damage was clearly a hallmark of aging. But was it the cause? Or just one of many fingerprints left behind?

By the time she published her first paper, Leena was no longer looking for a single villain. She had come to think of aging as a system failure—a layered collapse in which genomic instability played a starring, but not solitary, role.

And in that sense, her patients weren't just victims of radiation. They were early arrivals to a fate we all share: the gradual breakdown of the code that once made us young.

HOW OUR DNA TAKES DAMAGE

At cruising altitude, 35,000 feet above sea level, the sky changes. It becomes thinner, darker, and more electric. Commercial airline pilots sometimes speak of the "cosmic rain"—an invisible drizzle of particles that shower down from the far reaches of space, slipping through Earth's magnetic shield and colliding with atoms in the upper atmosphere. To the naked eye, there's nothing. But to a sensitive enough detector, the cabin hums with invisible strikes. These are high-energy particles—muons, protons, fragments of atomic nuclei—traveling at nearly the speed of light.

Radiation physicists know this phenomenon well. It's one reason pilots and flight attendants receive more ionizing radiation exposure annually than nuclear plant workers. The constant bombardment, while minuscule per flight, accumulates over thousands of airborne hours. DNA doesn't care if the ionizing charge came from a Chernobyl reactor or a solar flare—when it hits, it hits. A single energetic particle can punch through skin, muscle, even bone, before striking a strand of DNA with enough force to snap it.

There are photographs of this damage. In controlled lab settings, human cells exposed to low-level radiation show twisted chromosomes, shattered double helices, fragments that no longer align with their original genetic partners. In a normal environment, at sea level, most of us avoid the brunt of these cosmic intrusions. But even on the ground, life is never free of assault.

Evolution, for all its elegance, did not create a static body. It created a living one—a structure constantly maintained in spite of entropy. DNA, far from being protected in a vault, sits right at the intersection of the living world and its hazards. And so, every moment of life involves a delicate struggle to keep that code legible.

Most people imagine aging as a slowburning decline in organs or muscles—things we can see and feel. But long before the body changes, long before skin loosens or memory falters, the earliest signs of aging begin at the molecular level, in a quiet war between our cells and the environment. That war takes place in the genome.

Inside each of your cells is about two meters of DNA, coiled into 46 chromosomes and packed with nearly 20,000 genes. That DNA is copied, transcribed, and checked millions of times each day. But the conditions under which it operates are far from sterile. Think of a librarian forced to recopy the same manuscript by hand while surrounded by flickering candles, gusts of wind, and clumsy assistants who keep smudging the ink. That's cellular life. Messy. Loud. Dangerous.

The sources of damage are nearly endless.

Some come from outside—ultraviolet light from the sun, tobacco smoke, industrial pollutants, radiation from natural rocks in the earth, cosmic particles. Others come from within—reactive oxygen species generated during metabolism, spontaneous chemical changes, replication errors during cell division. Even the act of reading DNA to produce proteins can, under stress, cause certain segments to fold or fray.

What's remarkable isn't that damage occurs. What's remarkable is how much of it the body repairs. Enzymes like ATM, PARP, BRCA1, and dozens more form a microscopic triage team—identifying breaks, patching errors, excising incorrect bases, rejoining severed strands. In younger organisms, this system is breathtakingly efficient. It repairs as quickly as damage appears.

But with age, that efficiency wanes.

Some repair pathways slow down. Others become errorprone. In some cases, the repair itself becomes a source of mutation—rejoining the wrong DNA ends, or introducing gaps. These mutations—most of them called somatic, meaning they occur after birth and are not inherited—accumulate slowly. By the time you reach your 70s, your tissues carry tens of thousands of these molecular scars. And they are not evenly distributed. Rapidly dividing tissues—like the gut, skin, and blood—see the highest mutation loads. But even slowdividing tissues like muscle and brain are not spared. Every cell ages. The only difference is in how fast, and with what consequences.

What makes these mutations dangerous isn't just their quantity. It's their location. A single mutation in a stretch of so-called "junk DNA" might do nothing. But if the damage strikes a gene responsible for DNA repair, or cell cycle regulation, or apoptosis—the orderly death of a defective cell—the impact compounds. You don't just get more damage. You lose the tools to stop the damage.

The result is a cellular environment that becomes gradually more chaotic. Some cells stop dividing entirely, entering senescence. Others divide recklessly, becoming precancerous. Some mutate silently for years, neither dying nor misbehaving—until a final mutation tips them into dysfunction. All of this happens before the first gray hair.

For decades, biologists debated whether these random mutations could truly drive aging. Some argued they were passengers, not drivers—evidence of decline, but not its cause. But then came a series of experiments that shifted the conversation.

Mice engineered with defects in DNA repair genes aged rapidly. Their skin wrinkled early. Their bones thinned. Their stem cell pools depleted. Some developed cancers young; others simply collapsed from organ failure. Meanwhile, mice with enhanced DNA repair systems showed slower aging trajectories—fewer mutations, better tissue integrity, and longer lifespans. Even among long-lived species like bats and naked mole rats, researchers found unusually robust genome maintenance systems.

EASY ACCESS ***Why this experiment matters***

Take away a cell's ability to fix DNA, and it ages on a stopwatch. The ERCC1 mice are not a curiosity — they are proof that the connection between damaged DNA and a damaged body is causal, not just correlated. Damage that the body cannot repair becomes damage the body has to live around. Multiply that across decades, across trillions of cells, and you have the slow shape of human aging.

Still, the picture remained complicated. When scientists gave mice hyperactive versions of tumorsuppressor genes—making their cells too sensitive to damage—they avoided cancer but aged prematurely. Their cells stopped dividing too early. Their tissues failed to regenerate. It was as if the body had chosen safety over vitality.

That tradeoff—between genome surveillance and cellular renewal—is now understood as one of the core paradoxes of aging. You can protect your body from cancer by stopping damaged cells from dividing. But you pay a price: your tissues lose the ability to regenerate, and aging accelerates.

CLOSER LOOK ***How a cell decides whether to repair, retire, or self-destruct***

When a strand of DNA breaks, the first responders are sensor proteins — ATM and ATR — that detect the lesion within seconds. They phosphorylate a protein called p53, encoded by the gene TP53. Activated p53 functions as a molecular triage officer: it pauses the cell's division cycle to buy time for repair. If the damage is minor, repair enzymes fix it and the cell moves on. If the damage is severe, p53 commits the cell to senescence — permanent retirement — or to apoptosis, programmed self-destruction. This is the body's tradeoff between cancer and aging in plain view: the same machinery that prevents a damaged cell from becoming a tumor also depletes the pool of healthy cells available to renew tissue. TP53 is, in fact, the single most commonly mutated gene in human cancers — disabled, the bouncer is gone.

Vousden & Lane (2007). p53 in health and disease. Nature Reviews Molecular Cell Biology, 8(4), 275–283.

This isn't a flaw. It's an evolutionary compromise. In a world where most organisms die from infection, injury, or predation long before they die of old age, the genome only needed to last long enough to reproduce. After that, maintenance is optional.

EASY ACCESS *The vigilance trap*

A body that punishes every damaged cell never gets cancer. It also runs out of cells. The mice with hyperactive p53 prove the point — perfect cancer defense, premature old age. Aging, in this sense, is the negotiated truce between two bad outcomes: tolerate too much damage and you get tumors, tolerate too little and you get frailty. Most of us drift toward frailty, slowly, because evolution tuned the system for early-life vigor — not for the eighty extra years modern medicine gave us.

But in humans, who now live decades beyond their reproductive years, this compromise becomes painfully visible. The genome, so exquisitely built for earlylife vigor, begins to fail. Not suddenly. Not catastrophically. Just quietly, strand by strand.

That slow loss of fidelity—barely perceptible at first—is the signature of genomic instability. It's not dramatic. It doesn't announce itself. But over decades, it rewrites what it means to be alive.

WHEN REPAIR FAILS

Deep in the Atacama Desert of northern Chile—one of the driest places on Earth—there lives a plant that shouldn't be alive. Its name is Nicotiana attenuata, more poetically called the coyote tobacco plant. For most of the year, it lies dormant beneath the baked crust of the desert floor. It waits for what rarely comes: rain. And when it does, the plant springs to life with astonishing speed. It germinates, grows, flowers, and seeds—all within a few weeks—before the next stretch of drought sets in. What makes Nicotiana attenuata extraordinary, however, is not its ability to survive with almost no water. It's its ability to withstand damage—genetic damage—on a scale that would cripple other organisms.

During those short, violent bursts of life, the plant is bombarded by ultraviolet radiation. With so little cloud cover, the sun's rays strike the desert unfiltered. The heat alone would destabilize the DNA in most life forms. But Nicotiana attenuata possesses an unusually robust DNA repair system. It can take hits—hundreds of lesions per cell—and correct them with a speed and fidelity that rivals anything seen in mammalian biology. Researchers studying the plant found that its genome expresses a higherthannormal concentration of repair enzymes, particularly those involved in base excision repair (BER) and homologous recombination (HR)—pathways that patch up small and large DNA breaks, respectively. It's as if the plant evolved not just to live in damage, but to repair its way through it.

Human cells once had a similar grace.

In our early years, DNA repair is swift and seamless. Cells detect double-strand breaks and immediately respond—through mechanisms like non-homologous end joining (NHEJ) or homologous recombination—often fixing the damage before the cell even proceeds to its next division. Oxidized bases are clipped out and replaced. Mismatched nucleotides are detected and corrected. These repair systems act like elite response teams: precise, well-resourced, and coordinated.

CLOSER LOOK ***Two ways a cell repairs a broken strand — and why one is better at it***

A double-strand break is the most dangerous kind of DNA damage: both rails of the helix are severed at once. Cells use two main strategies to fix it. Homologous recombination (HR) waits until the cell has already copied its DNA, then uses the intact sister copy as a template. The repair is essentially perfect. Non-homologous end joining (NHEJ) is the rougher alternative: the cell simply ligates the loose ends back together, sometimes losing or adding a few base pairs in the process. NHEJ is fast, available all the time, and the workhorse of most break repair — but it is error-prone. With age, the balance between the two systems shifts, and the fraction of breaks repaired imperfectly creeps upward. The genome, slowly, becomes a patchwork.

Niedernhofer et al. (2018). Nuclear genomic instability and aging. Annual Review of Biochemistry.

But they age.

With time, the efficiency of DNA repair declines. Not all at once, and not in all tissues equally. But the effect is cumulative. The proteins involved in repair—many of which require complex multi-step assemblies—become less reliable. The signaling that initiates repair weakens. The enzymes themselves are sometimes misfolded or mutated. And in stem cells, which rely on pristine genomes to produce healthy progeny, the cracks begin to show first.

Stem cells are the body's most long-lived workers. They persist for decades, maintaining tissues by dividing into specialized cells while preserving their own population through self-renewal. But that renewal depends on accuracy. A stem cell carrying too many mutations becomes a liability—it can seed malfunctioning tissue or ignite tumors. So when damage accumulates beyond a threshold, the body makes a choice. It shuts the stem cell down.

This is not a gentle retirement. The cell is forced into senescence—a state of permanent arrest where it can no longer divide. Or it undergoes apoptosis: a clean, programmed self-destruction. From the body's perspective, this is risk management. Better to lose a cell than allow it to turn rogue. But over time, the loss adds up.

Every tissue that depends on stem cells begins to thin. The lining of the gut, once renewed every few days, regenerates more slowly. Skin heals less efficiently. Bone marrow, which produces blood, starts to falter. And in organs like the brain, where stem cell turnover is rare but vital, the signs appear as cognitive dulling or motor decline.

This isn't theoretical. Scientists have created mouse models where specific DNA repair pathways are genetically disabled. In one such model, mice lacking a critical repair gene called ERCC1 aged rapidly. They developed osteoporosis, muscle wasting, and neurological deficits within months. Their organs failed not because of external injury, but because the cells within them stopped functioning. In another study, mice with disabled HR pathways accumulated chromosomal aberrations that led to early mortality. Their tissues—especially those with high cellular turnover—degraded visibly and quickly.

These findings are not just dramatic. They're causative.

Remove the repair, and aging accelerates.

AGING, CANCER, AND THE PRICE OF PROTECTION

In 1971, President Nixon declared war on cancer. Billions of dollars followed—funding laboratories, equipment, research trials. What followed was one of the most ambitious scientific efforts in modern history: an attempt to defeat a disease that wasn't a foreign invader, but something generated from within.

What researchers didn't fully appreciate at the time was that cancer wasn't a disease in the traditional sense. It was a consequence—a downstream outcome of a deeper process. And that process, as decades of work would eventually reveal, begins with genomic instability.

We tend to think of cancer as an anomaly. A mistake. But in truth, it's an evolutionary response gone too far. When cells replicate, they make errors. That's unavoidable. When those errors affect key genes—those involved in division, death, or DNA repair—the result can be a cell that multiplies uncontrollably. But the body doesn't simply allow that to happen. It has evolved layers of defenses, mechanisms that monitor DNA for irregularities and halt proliferation when thresholds are crossed. These defenses, however, operate under constraints. They can't shut down every flawed cell without sacrificing tissue function. And so, a negotiation takes place. Some cells are eliminated. Others, damaged but not disastrous, are allowed to persist.

With age, the terms of that negotiation change.

Every tissue in the body carries its own vulnerability profile. The skin, constantly exposed to UV radiation, accumulates signature mutations—C to T transitions at dipyrimidine sites—that reveal the scars of sun exposure. The colon, with its high turnover and rich microbial environment, sees mutations associated with oxidative stress and inflammation. The lungs absorb carcinogens in air. The liver processes toxins in food and drink. Each organ tells a different story, but the script is the same: damage, imperfect repair, accumulation.

In healthy young individuals, many of these mutations are tolerated. They sit quietly, having landed in non-coding regions or in genes where a single change does little harm. But over time, and especially with repeated exposure, the chances of hitting a critical gene increase. These aren't mutations in junk territory. They land in oncogenes—which, when activated, tell a cell to divide. Or in tumor suppressor genes, which normally act as brakes. Disable one, and the cell becomes less cautious. Disable both, and it may no longer respond to signals that would otherwise tell it to die.

These are the seeds of cancer. And they are far more common in the aging body than most realize.

Recent advances in sequencing have made it possible to map the genomes of cells taken from ordinary, non-cancerous tissues. The findings are sobering. Skin cells from elderly individuals carry thousands of somatic mutations per cell. Blood samples from older adults often contain clonal populations—groups of cells descended from a single mutant ancestor, each bearing the same driver mutation. This phenomenon, known as clonal hematopoiesis, is now understood as a common feature of aging. It's not leukemia. But it's one step closer.

The presence of these mutations isn't enough to cause disease. Most never will. But they shift the landscape. They raise the baseline level of risk. Aging, in this view, is less about time and more about probability. The longer a tissue exists, the more mutations it accumulates. The more mutations it accumulates, the more likely it is that one—or several—will push a cell over the edge.

And yet, the body doesn't give in easily.

Even damaged cells can be controlled, provided their internal circuitry still responds to regulation. That's where genes like TP53—encoding the p53 protein—play a central role. When DNA damage is detected, p53 is activated. It halts the cell cycle, buys time for repair, and, if necessary, triggers self-destruction. In essence, p53 is a bouncer at the molecular level. And for much of your life, it works well.

But like all systems, p53 has its limits. Sometimes it is mutated itself, neutralized by the very cells it's supposed to guard against. In fact, TP53 is the single most commonly mutated gene in human cancers. When it fails, the body's last line of defense is gone.

Even when it functions, p53 is not without cost. In aging tissues, especially those where stem cell pools are finite, frequent activation of p53 contributes to cell cycle arrest and depletion of regenerative capacity. It keeps the tissue safe from tumors—but unable to renew. This is the price of protection. You get cancer prevention at the expense of tissue vitality. The body chooses caution over performance.

And so, the genome—imperfect, mutable, dynamic—drifts further from its original state. By the eighth or ninth decade of life, the average human body is a mosaic of genetic diversity. No two liver cells are the same. No two skin cells carry the same code. Your body, once derived from a single zygote, becomes a crowd of slightly miscopied instructions.

Not all of this leads to disease. Much of it is neutral. Some of it is adaptive. But some—just enough—opens the door to dysfunction.

The most vivid expression of this dysfunction is cancer. But there are quieter versions. A failing hematopoietic system that produces malformed blood cells. A retina where a single mutation causes progressive degeneration. A brain in which somatic mutations disturb local circuitry, contributing to decline.

Genomic instability doesn't just trigger disease. It creates variability. And with variability comes risk.

Still, one question remains: if these mutations are so widespread, why do some people live to 100 with relatively intact health, while others decline at 60?

The answer, in part, lies not in the damage itself, but in how the body responds. Surveillance, repair, redundancy—all of these play roles. Some people inherit more efficient DNA repair pathways. Some have immune systems better equipped to cull damaged cells. Some are simply lucky—their mutations land in regions that don't matter.

But even the lucky don't escape entirely. The blueprint degrades for everyone. What differs is how fast, and how well the building can compensate.

THE PARADOX OF PERFECTION — WHEN DNA REPAIR BECOMES TOO GOOD

In the high Arctic, the snowy owl thrives in a world of extremes. With piercing yellow eyes and feathers as silent as snowfall, it hunts over frozen tundra where life is scarce and mercy scarcer. Everything about this bird seems engineered for precision. It sees ultraviolet patterns invisible to its prey. It hears the scurry of rodents under a foot of snow. And it times its breeding not by temperature, but by the population cycles of lemmings—its primary food source. If lemmings are abundant, the owl will raise a brood of ten. If not, it won't lay a single egg. No waste. No excess. Survival hinges on restraint.

Biologists studying the snowy owl's reproductive strategy have long puzzled over its unforgiving minimalism. Other birds hedge bets, raise fewer chicks, or adapt gradually to resource shortages. But the snowy owl operates differently.

It withholds completely unless the environment is just right. This rigidity is its strength—and also its risk. Because in a year of miscalculation, when the lemming count is misjudged, the owl has no fallback. It breeds too little. And the population suffers.

Nature is full of systems that favor strict control—tight thresholds, failsafes, rigid logic. But in biology, as in the Arctic, perfection comes at a price. The same principle plays out not only in snowy owls, but in your cells.

In a landmark study, researchers created transgenic mice with a hyperresponsive form of p53. These mice were remarkably resistant to tumors. They rarely developed malignancies, even into old age. On paper, it looked like a victory—until the mice began to deteriorate. Their bones thinned early. Their skin lost elasticity. Their stem cell reserves declined rapidly. They developed kyphosis, a stooping curvature of the spine, and died young. The system had been protected from cancer. But in the process, it had burned through its regenerative capacity.

What these experiments revealed was a counterintuitive truth: too much DNA repair can be harmful. Or more precisely, too much DNA surveillance, coupled with zero tolerance for damage, can undermine tissue maintenance. The logic is cruel but simple. Every time a cell with minor damage is permanently shut down, the organ loses a little function. Do that often enough, and the organ weakens—not from disease, but from attrition.

This phenomenon is not unique to mice. In human aging, we see a similar dynamic. As tissues accumulate DNA damage, p53 and related pathways become more active. This rising vigilance might protect against latelife cancers, but it also forces cells into senescence. Senescent cells are alive but inert. They occupy space, emit inflammatory signals, and disrupt their environment. They don't divide. They don't contribute. They linger.

Evolution has long favored organisms that can defend against early threats. Cancer is one of those threats. A single malignant cell, left unchecked, can destroy a host. Aging, by contrast, is slow. It unfolds over decades, often after reproduction. Natural selection cares more about the sprint than the marathon. And so the systems that evolved to guard our genomes were designed with short-term survival in mind.

This explains why organisms with extreme cancer resistance often exhibit signs of accelerated aging. They've tilted the balance toward strict enforcement. Some of the longestlived species, like elephants, appear to have resolved this tradeoff differently. Elephants carry 20 copies of the TP53 gene, compared to humans' one. But they also have regulatory mechanisms that prevent hyperreactivity. Their p53 system is buffered, allowing for repair without overreaction. In contrast, engineered systems—like those in lab mice—lack this nuance. They show us what happens when the pendulum swings too far.

CLOSER LOOK ***Twenty copies of the guardian — what elephants taught us***

Cancer risk should scale with body size and lifespan: more cells, more divisions, more chances for a mutation to land in the wrong place. Elephants, with roughly a hundred times the cells of a human and lifespans approaching seventy years, should be cancer factories. They aren't. Estimated cancer mortality in elephants is under five percent, compared to more than twenty percent in humans. The reason, at least in part, is remarkable: elephants carry twenty copies of the TP53 gene. Humans carry one. When elephant cells take DNA damage, their amplified p53 system responds faster and more decisively, pushing damaged cells into apoptosis before they can become tumors. The lesson isn't that we should engineer more p53 — when scientists tried that in mice, the animals avoided cancer but aged prematurely. The lesson is that evolution has already discovered the tradeoff between vigilance and renewal, and solved it differently in different species.

Abegglen et al. (2015). Potential mechanisms for cancer resistance in elephants. JAMA, 314(17), 1850–1860.

The analogy with the snowy owl became strangely apt. An organism that insists on optimal conditions may succeed when the world cooperates—but when the environment falters, that same rigidity turns fatal. The human genome, like the owl's breeding instincts, needs flexibility to endure. A cell that forgives small damage may function longer, even if it risks occasional error. A cell that panics at every lesion may retire too soon.

Aging, viewed through this lens, is not merely damage accumulated or function lost. It is a negotiation between vigilance and vitality. Between the need to guard the blueprint and the need to keep building.

And sometimes, perfection is the wrong goal. Is the Blueprint the Villain — Or Just the Beginning of the Story?

In 2016, researchers at Harvard did something unsettling. They took mice that were visibly aged—stooped, grayfurred, slowmoving—and subjected them to a genetic intervention. But they didn't repair the mice's DNA. They didn't remove any mutations or replace damaged sequences. Instead, they altered the way the DNA was read. Specifically, they reset the epigenetic markers—the molecular instructions that determine how genes are turned on or off.

CLOSER LOOK ***Resetting the reader without rewriting the book***

The 2016 Salk experiment used a technique called partial cellular reprogramming. Researchers transiently expressed four genes — Oct4, Sox2, Klf4, and c-Myc, collectively known as the Yamanaka factors — in living mice. Sustained expression of these factors converts a mature cell back into a stem cell, which is too dangerous in a whole organism: it triggers tumors. But cyclic, short pulses do something subtler. They wipe age-related epigenetic marks — the chemical tags on DNA and its packaging proteins that tell each cell which genes to switch on — without erasing the cell's identity. The DNA itself was untouched. The mutations were still there. What changed was how the genome was being read. Aged tissues recovered function. The progeroid mice in the study lived thirty percent longer than untreated controls. The implication, still being worked out, is that part of what we call aging is not damage to the score — it is the orchestra forgetting how to play it.

Ocampo et al. (2016). In vivo amelioration of age-associated hallmarks by partial reprogramming. Cell, 167(7), 1719–1733.

Within weeks, the mice began to change. Their muscles strengthened. Their eyesight improved. Their behavior became more active. To anyone observing, it looked like the clock had been turned backward.

What made the experiment provocative wasn't just that it worked. It was that it worked without correcting the underlying mutations. The DNA remained as damaged as before—scarred by time, marked by the same point mutations and structural breaks. Yet the animals regained function. Not perfectly, not completely—but enough to question what many had assumed.

If the genome was broken, how was it that the body still remembered youth?

This experiment, and others like it, shook a basic belief in the field of aging: that genomic instability—the accumulation of damage to the DNA itself—was the singular engine of decline. The results didn't deny that damage was present. They simply asked whether damage, by itself, was enough to cause aging.

Leena Patel had once thought so. Her patients, many of whom had been exposed to environmental radiation, bore the marks of accelerated mutation. Their tissues collapsed.

Their stem cells depleted. The evidence felt clear. DNA damage was the trigger, and everything else followed.

But even she had begun to feel the boundaries of that explanation. Some patients, despite high mutation burdens, functioned well into old age. Others declined rapidly with fewer obvious defects. There were patterns, but there were also exceptions. The genome was telling a story, but it wasn't telling the whole one.

The blueprint metaphor, once helpful, began to feel incomplete.

A blueprint is passive. It sits still, awaiting interpretation. But the genome doesn't work that way. It's read constantly, translated, modified, silenced, and reshaped depending on cellular needs. One cell can read the same stretch of DNA differently than another. And those interpretations are controlled not by the sequence alone, but by the epigenetic landscape layered on top—chemical tags, methylation patterns, histone configurations.

In this view, genomic instability remains a hallmark of aging—but not the only one. It is part of a cascade, a deterioration that includes not just the damage, but the cell's capacity to interpret damage, repair it, or route around it. In some cases, the mutations accumulate harmlessly. In others, the cell spirals into dysfunction—not because of what was written in the code, but because of how the code was read.

This subtle shift in thinking changes the implications. It suggests that the genome is not a villain. It's a witness. A record of every insult, every repair, every failure to correct. Over time, it becomes harder to tell the difference between what is essential and what is detritus. The fidelity of the code matters, but so does the flexibility of the reader.

That flexibility degrades with age. Cells lose the ability to prioritize. They activate stress responses when none are needed. They silence genes that once protected them. They forget which pathways lead to repair and which to destruction. It is this loss of responsiveness, more than the damage itself, that may mark the true beginning of decline.

Still, the presence of damage matters. Even if it can be ignored or compensated for in the short term, the accumulation of mutations—especially in stem cells—shapes the landscape of aging. It defines what is possible, what is repairable, what is inevitable.

And sometimes, that damage cannot be bypassed.

Cancer is the most visible result. But there are subtler failures. A cell that stops dividing in the gut lining. A misfiring neuron in the prefrontal cortex. A retinal cell that can no longer maintain its shape. These failures are not catastrophic on their own. But over time, they produce the long, slow disintegration we call aging.

The body's surveillance systems—p53, ATM, BRCA1, and the rest—respond with the tools they have. They repair where they can. They sacrifice where they must. And they err, constantly. Sometimes they overcorrect. Sometimes they miss the warning signs. The net result is a body that, over the decades, shifts from stability to drift.

The blueprint still exists. But it's no longer followed with precision.

By the end of her decade-long study, Leena had come to see the genome differently. She no longer imagined it as a book whose pages had been torn. She saw it more as an evolving script—constantly rewritten, adapted, sometimes misread, occasionally redacted altogether. What mattered, she realized, wasn't whether the script was perfect. It was whether the actors could still perform the play.

Some could. Some couldn't. And in that divergence—in that variability between cells, tissues, and people—lay the real story of aging.

Genomic instability remains a foundation. It sets the terms. But the house we build on top of it—the function, the repair, the resilience—that's where aging is won or lost.

We age not simply because the blueprint breaks. We age because the system forgets how to read it.

THE CLOCK AT THE ENDS OF CHROMOSOMES

TELOMERE ATTRITION

It wasn't the anemia that caught Dr. Saira Mahmud's attention. In a community clinic nestled among the wheat fields of rural Punjab, anemia was practically I routine. Children arrived pale and listless; adults bore it with resignation. But what puzzled her were the siblings—Ayaan and Zoya—both in their twenties, both presenting with bodies that told a much older story.

Zoya's hands were creased and rough like an old seamstress. Her hair, once jetblack, had grayed at the temples. She complained of breathlessness climbing stairs, chronic mouth ulcers, and what she called "spells of weakness." Her younger brother, Ayaan, had more alarming symptoms: chronic fatigue, patches of skin that didn't heal, and a white blood cell count so low she initially thought it was a lab error. But it wasn't. His immune system had all but stalled.

Neither smoked. Neither had chronic illnesses. Their medical history was, until recently, unremarkable. Yet here they were—slumped on her examination table with marrow that struggled to produce blood, immune cells that refused to regenerate, and bodies collapsing under invisible weight. Zoya's lung X-rays showed early signs of fibrosis, though she'd never worked near dust or chemicals. Ayaan's bone marrow biopsy came back marked by hypocellularity. It was like watching two people unravel from the inside out.

At first, she thought it might be a rare form of aplastic anemia or a mitochondrial disorder. She combed through their family tree, looking for clues—nothing obvious. No consanguinity, no inherited metabolic diseases, no early deaths in close relatives. The parents were poor, but healthy. A few aunts with arthritis, a grandfather who died of tuberculosis. Nothing that would explain why two otherwise healthy young adults were aging like war veterans.

She reached out to a colleague in Karachi, a hematologist who'd seen something similar once, years ago, in a military hospital. "Dyskeratosis congenita," he suggested over a scratchy phone line. "It's rare, but when it shows up, it does this—rapid aging, marrow failure, nail dystrophy, skin pigmentation. Very short telomeres."

Telomeres. That was the word that began to change everything.

Saira had heard the term in medical school—briefly. Something to do with the ends of chromosomes. Protective caps, she thought. Like shoelace tips? Or was that mitochondria? She wasn't sure. But the idea that a defect in these invisible genetic endcaps could unravel entire organ systems seemed almost poetic. And horrifying.

She ordered genetic tests—an expensive and bureaucratic process for her underfunded clinic. Samples were couriered to a lab overseas, and she waited. The patients deteriorated slowly as summer wore on. Zoya began to lose weight. Ayaan developed gum bleeding and a strange cough that lingered like smoke. They started wearing masks before COVID made it a norm—not for prevention, but because every sneeze in the room felt like a loaded gun.

The results arrived in mid-August. The diagnosis was unequivocal. Both siblings had mutations in the DKC1 gene—a hallmark of dyskeratosis congenita. Their telomeres, the lab reported, were shorter than those typically seen in 70-yearolds.

Saira stared at the report in silence. She wasn't just treating a disease. She was staring at a mechanism—a cellular clock, it seemed, that had run out decades too soon. It was the first time she grasped that aging, so often described as natural, could begin with a precise defect. That there were individuals whose aging wasn't metaphorical—it was molecular. And mercilessly fast.

There was no cure. Treatment options were sparse. Bone marrow transplants had been tried in some cases, with mixed results. Experimental therapies existed in clinical trials, but her patients would never qualify—too remote, too poor, too forgotten by the systems that made such miracles possible.

She sat with Ayaan and Zoya and explained it as gently as she could. She didn't speak in metaphors. There was no "fountain of youth" to contrast against. Just the truth: that something deep inside their cells—the very thing meant to guard their genetic material—was fading with every passing month.

They took it better than she did. Zoya nodded quietly. Ayaan cracked a joke about finally having a diagnosis fancy enough to write a memoir. But that night, after they left, Saira sat in the dispensary and re-read every page of the report. The word \"telomere\" kept appearing like a whisper at the end of every sentence.

It would be weeks before she found time to research it properly. Before she uncovered the story of how scientists discovered these repeating DNA sequences. How they unraveled a mystery at the ends of chromosomes that echoed all the way into the cells of her two young patients.

EASY ACCESS ***What telomeres are, in one paragraph***

Every chromosome ends in a long stretch of repeated DNA letters — TTAGGG, written hundreds of times. These caps don't code for anything. They exist to be sacrificed. Each time a cell divides, the copying machinery loses a little material from the ends, so the telomere shortens by a few dozen letters. As long as the cap is long enough, no important genes get clipped. When it gets too short, the cell stops dividing. That is the essence of the telomere clock — not damage, but planned, predictable wear in a buffer zone built for exactly that purpose.

But in that moment—staring at two people who were not dying, just aging far too fast—Saira knew she had stumbled on something much larger than a rare disorder.

THE DISCOVERY OF THE SHOELACE ENDS

In the late 1970s, long before the phrase "anti-aging" flooded supermarket shelves, a quiet revolution was underway in the cool, humming corridors of academic labs. At Yale, and later at the University of California, a young biologist named

Elizabeth Blackburn was staring at something strange—something small and repetitive at the edges of chromosomes that didn't seem to belong.

She wasn't trying to solve the mystery of aging. She was trying to understand how singlecelled organisms kept their chromosomes intact. It started with a species of ponddwelling protozoa called Tetrahymena, whose chromosomes had a curious habit: they ended in the same short DNA sequence, repeated over and over again like the tail end of a song stuck on loop. Not once or twice, but hundreds of times.

That sequence—TTAGGG—didn't code for any protein. It was non-coding, repetitive, and utterly consistent across all the chromosome ends she examined. At first, it seemed like an evolutionary accident, some leftover noise in the genome. But Blackburn had a different instinct. She suspected it served a purpose.

By the early 1980s, she and her collaborator Jack Szostak had uncovered the truth: these repeating segments acted as protective caps. They weren't junk. They were shields. Every time a cell divided, its machinery could replicate the bulk of the DNA just fine—but it always struggled with the ends. The copying process simply couldn't finish the job. A little bit of genetic material was lost with each duplication. But instead of shaving off vital genes, the cell lost these end repeats—telomeres. And that, it turned out, was by design.

Without them, chromosomes frayed. They stuck to one another. They lost structural integrity, like ropes unraveling at the ends. But with telomeres in place, the genome held together.

The shoelace analogy came later—those plastic aglets at the ends of laces that keep the fibers from fraying. But Blackburn's discovery didn't need a metaphor to make waves. It fundamentally changed how biologists thought about cellular replication. The ends weren't dead zones. They were sacrificial buffers. And they wore down with every generation.

Still, this raised a sharper question. If telomeres short-ened with each division, why didn't cells just degrade after a few generations? Why weren't we all puddles of degraded DNA by the time we were toddlers?

That's when Blackburn, along with Carol Greider—a PhD student with a knack for methodical obsession—discovered the enzyme that would complete the picture: telomerase.

They found it in Tetrahymena, too. It was an odd little protein with its own built-in RNA guide, like a molecular mechanic reading blueprints as it worked. Telomerase did what the cell's usual replication tools couldn't: it rebuilt the chromosome ends. Slowly, precisely, it extended the telomeres—adding more of that same repetitive sequence back to the ends that were wearing thin.

It didn't operate in every cell. Most adult cells never saw it. But in the germline—egg and sperm—telomerase was active, keeping chromosomes pristine for the next generation. It also showed up in stem cells, those rare, self-renewing reservoirs that supported tissue repair. Telomerase, it turned out, wasn't just a quirky enzyme. It was a key part of the maintenance crew. A way for certain cells to reset the clock.

The discovery was met with awe and a dash of unease. For the first time, scientists could see a built-in limit to how many times a normal cell could divide. And they had found the enzyme that could extend that limit. Greider would later say the enzyme's name came quickly—"telomerase" was both literal and intuitive—but its implications were anything but simple.

The idea of a molecular clock wasn't new. But this was something else. This was a mechanism, not just a metaphor. A biological countdown embedded in every dividing cell, with a tool that, if reactivated, could potentially stop or reverse it.

For a while, telomeres were a curiosity of molecular biology—an elegant solution to a replication problem, more relevant to yeast and protozoa than to people. But that changed as researchers began checking human chromosomes. The same TTAGGG repeats were there, lining our own chromosome tips like invisible countdown timers. Human telomerase was harder to find—but it was there too, operating mostly in stem cells, embryonic cells, and notably... in tumors.

Before the 1990s ended, telomerase had gone from an obscure enzyme in pond sludge to a molecule of intense interest in aging, cancer, and biotechnology. It carried the mystique of power—not because it promised youth, but because it offered precision. Unlike haphazard genetic fixes, telomerase operated on a simple truth: keep the ends intact, and you buy the cell more time.

In 2009, Blackburn, Greider, and Szostak received the Nobel Prize for their work. It was one of those rare moments when molecular biology crossed into public consciousness. Telomeres were suddenly on magazine covers. Telomerase was whispered about in the longevity community. But the real legacy of their discovery wasn't in miracle creams or popscience headlines. It was in the clarity it brought.

Cells age. And part of that aging, it turns out, is written not in scars or stress or oxidative chaos—but in length. In silent, repeating letters at the edge of every chromosome.

They do not scream. They do not warn. But they do run out.

THE CELLULAR CLOCK STARTS TICKING

Inside every somatic cell—the skin cell shielding you from sunlight, the lining of your blood vessels, the fibroblast waiting to heal a wound—there is a quiet transaction under-way. It begins the moment the cell divides. A copy of the genome must be made. Fortysix chromosomes, billions of base pairs. The machinery is exquisite, but imperfect. Not in accuracy, but in architecture.

There's a quirk in the way DNA is replicated. A mechanical oversight left behind by evolution, tolerated rather than corrected. Each time the cell copies its DNA, it loses a little bit off the end of each strand. This isn't because the cell is sloppy. It's because the enzymes responsible for replication—the polymerases—require a small piece of RNA to start their work. Once the copying begins, the machinery races forward, building a perfect twin of the chromosome. But when it reaches the very end, it simply stops. There's no room to anchor. No place to finish the job.

The result is a few missing letters at the tail of every chromosome. Not once. Every time.

Biologists once wondered why this didn't immediately cripple the genome. Losing genetic material sounds catastrophic. But nature, anticipating this problem, buffered the system. It placed long, repetitive, non-coding sequences at the ends of chromosomes. These were the telomeres. Not genes. Not blueprints. Just filler—sequences that could be trimmed without consequence. At first.

The analogy that eventually stuck was that of a photocopier. Imagine copying a book over and over, but the machine always clips the last few words of the last sentence on every page. You keep copying the new copy. Over time, even if the story remains legible, the endings blur. Eventually, the loss reaches the substance. The plot begins to vanish. The telomeres are those final words—the sacrificial lines that take the hit so the rest of the text stays intact.

But unlike paper, the cell doesn't allow its core content to erode. It has a defense mechanism. A limit.

At a certain point, when the telomeres become too short, the cell senses danger. The exposed ends of chromosomes begin to resemble DNA breaks—signals of damage that must be contained. The cell panics. But instead of repairing or continuing blindly, it triggers a shutdown.

Some cells enter senescence, a permanent state of non-division. They stay metabolically active, sometimes even perform basic functions, but they will never divide again. Others choose apoptosis—programmed cell death—a clean, regulated self-destruction.

This isn't a malfunction. It's a form of biological prudence. Cells are given a finite number of safe divisions, and when that number is up, they step aside. This limit was first observed in the 1960s by Leonard Hayflick, who noticed that human fibroblasts grown in a dish would divide only about 50 times before entering senescence. The phenomenon became known as the Hayflick limit, and it was one of the first cracks in the theory that cells were indefinitely renewable.

For years, scientists didn't know why the cells stopped. There was no toxin. No infection. Nothing obviously wrong. Just a refusal to divide. Telomeres provided the answer. With every division, they grew shorter. And once they dropped below a critical threshold, the cell recognized that it could no longer divide safely.

This clock wasn't just passive wear and tear. It was embedded in the design.

Unlike other forms of molecular damage—oxidative stress, protein misfolding, or mitochondrial dysfunction—telomere attrition was mathematical. Predictable. A countdown. The more a cell divided, the closer it came to its end.

And it wasn't reversible.

This countdown didn't mean the cell was "old" in any narrative sense. A cell could look healthy, function normally, and still be on its last few divisions. It was like a car with a full tank of gas but a wornout engine. One more long drive, and it might seize.

Interestingly, telomere length varied at birth. Some people started life with longer telomeres than others, giving them a wider margin. But once adulthood began, the shortening was consistent in most somatic cells. A few base pairs lost every time. Division after division. Decade after decade.

This process didn't require disease. It didn't need injury. It happened in healthy tissue, in healthy people, throughout their lives. The loss was microscopic but cumulative, invisible but absolute.

Telomere attrition wasn't an emergency. It was a memo: Proceed with caution. Your limit is approaching.

The precision of the system had its logic. By limiting the number of times a cell could divide, the body minimized the chance of runaway mutations. The fewer the replications, the fewer the errors. It was another evolutionary compromise—trading the dream of infinite renewal for the assurance of genomic fidelity.

But the cost was clear. Every division was borrowed time. Every healing wound, every immune response, every moment of growth pushed the clock forward. Eventually, for every somatic cell, the instructions would come: stop.

And so, aging begins not with pain or fatigue, but with a quiet pause deep inside the cell. A silence where once there was division. A countdown reaching zero.

WHERE IT HURTS THE MOST

At the heart of this selective vulnerability lay the quiet work of telomeres.

Bone marrow is one of the most demanding tissues in the body. Every day, it produces billions of blood cells—red cells to ferry oxygen, white cells to fight infection, platelets to clot wounds. The demand never stops. The marrow is a cellular factory with no offswitch, and its workforce is built on division. When telomeres shorten in these progenitor cells, the consequences are immediate and visible. Blood production falters. Immunity weakens. Anemia creeps in. The body begins to lose its frontline defenses.

In older adults, short telomeres in blood cells have been linked to a syndrome known as clonal hematopoiesis. It's not cancer—but it's close. A few blood stem cells begin to dominate production, crowding out diversity. The immune system becomes brittle, less adaptive, less responsive. Infections that once passed unnoticed become lethal. Vaccines become less effective. The body forgets how to remember its enemies.

Then there's the gut—a tube of muscle and mucosa lined with epithelial cells that are replaced every few days. This relentless turnover is essential. Every meal, every microbe, every swallowed toxin assaults this lining. Without constant renewal, the gut wall thins, absorption falters, inflammation festers. And as telomeres shorten, the regeneration slows. Microscopic tears stay torn longer. Barrier function deteriorates. The gut leaks—first figuratively, then literally.

The skin tells a similar story. It regenerates from the bottom up, with new cells pushing older ones outward in a slow cascade. In youth, this process is seamless. Cuts scab and vanish. Bruises bloom and fade. But with age, the outer layers lose their vigor. Wounds linger. Collagen breaks down faster than it is replaced. Telomere attrition in skin cells translates into fragility. Not just cosmetic aging, but structural vulnerability.

Even hair, often dismissed as vanity, is part of this story. Hair follicles rely on rapid division to produce new shafts. As the telomere clock runs down, follicles shrink. Growth slows. Pigmentproducing cells give out. The visible signs—thinning hair, gray strands—are more than aesthetics. They're signals of cellular exhaustion.

And perhaps most quietly but critically, the immune system bears the cost.

Lymphocytes, especially T-cells, divide rapidly in response to threats. But each division brings them closer to the telomere wall. In youth, the immune system mounts coordinated, powerful responses. It adapts, remembers, recalibrates. But with age—and with shortening telomeres—the memory fades. The T-cell repertoire narrows. Old threats return unchallenged. New ones go unnoticed. Chronic infections like cytomegalovirus creep back in. Inflammation simmers low-grade and constant, a phenomenon known as "inflammaging."

The observational data mirrors the biology. In large population studies, people with shorter telomeres in their white blood cells are more likely to show signs of frailty, immune decline, and poor response to stressors. They're more vulnerable to pneumonia, to sepsis, to complications from surgery. They recover more slowly. They age less gracefully.

BORROWED TIME

In the sterile hush of a 1950s Baltimore laboratory, a Petri dish pulsed with life that refused to end. Technicians watched in quiet disbelief as the cells multiplied, again and again, long after they were supposed to stop. Days turned to weeks. Weeks to months. Then years. What they didn't know—what no one could know yet—was that they were witnessing one of biology's most unsettling anomalies. The cells were from a woman named Henrietta Lacks, a cervical cancer patient at Johns Hopkins. The cells—now famously known as HeLa—had slipped the leash of cellular aging. And they had done it using a molecule very few people had heard of at the time: telomerase.

EASY ACCESS *Why HeLa cells matter to this chapter*

HeLa cells are the most famous cell line in biology — descendants of a cervical tumor taken from Henrietta Lacks in 1951, still dividing in laboratories around the world. They matter here because they prove the principle. Tumor cells reactivate telomerase. With telomerase running, the Hayflick limit goes away. With the limit gone, those cells will divide forever, given food and warmth. Mortality, in cells, is not a fixed feature of life. It is a switch that most adult tissues choose to leave off.

Before HeLa cells ever entered public imagination, telomerase was already at work in a different context: embryos. In the earliest stages of human life, every cell is dividing rapidly, building the foundation of an entire body from a single fertilized egg. During this period, telomerase is fully active. It constantly replenishes the ends of chromosomes, preventing telomeres from shortening. There is no countdown in the womb. Just expansion. Creation without constraint.

As development progresses, something shifts. In most tissues, telomerase expression falls silent. It doesn't disappear entirely, but it becomes highly restricted. Only a few cells retain it—germ cells, some immune cells during intense activation, and a handful of long-lived progenitor populations. Everywhere else, the silence is intentional. Evolution, it seems, prefers a built-in limit.

But telomerase is still there, encoded in our DNA like a forgotten key. It's a reverse transcriptase enzyme—a molecular craftsman that uses its own RNA template to rebuild the TTAGGG repeats at the chromosome ends. It doesn't just slow the shortening. It restores length. When active, telomerase essentially resets the clock.

Researchers have watched this in cell culture. Take a line of human fibroblasts—cells that normally senesce after fifty divisions. Introduce telomerase artificially, and those same cells keep going. They bypass the Hayflick limit. They divide a hundred times. Then two hundred. Then indefinitely. They don't die. They don't stop. They become, in the most literal sense, immortal.

It was this discovery that shifted telomerase from a niche curiosity to a molecule of profound interest. Because if you could reactivate telomerase in the right cells, under the right conditions, you might be able to reverse part of the aging process. At least at the cellular level.

But before the promise, there was awe. Telomerase wasn't just a longevity tool. It was a reminder of the strange exceptions life allowed. In species like hydra—those freshwater creatures that can regenerate indefinitely—telomerase remains active throughout life. In planarians, flatworms that regrow entire bodies from tiny fragments, telomerase hums continuously in their stem cell-like neoblasts. These animals don't age in the traditional sense. Their cells don't run out of time. They divide, renew, and rebuild indefinitely. And telomerase, always on, is part of that loop.

Even in the natural lifespan of humans, there are quiet exceptions. Male germ cells continue to divide across decades. Yet they don't show the same telomere erosion as somatic cells. Why? Because telomerase is still expressed. The reproductive line must remain intact, generation after generation. So the enzyme persists—not for the sake of the body, but for the blueprint it carries.

There's something haunting about that asymmetry. Your skin cells wear out. Your neurons don't replenish. Your immune system forgets. But your sperm—or the egg cells you began life with—preserve the code. They are designed to be ageless, even as the rest of you erodes. They don't age because they're not meant to live your life. They're meant to start someone else's.

That principle—longevity in service of continuity—runs deep in biology. Telomerase isn't a promise of personal immortality. It's a mechanism to safeguard lineage. Evolution didn't build it to keep us young. It built it to keep the species going.

And yet, the very fact that this enzyme still exists in our genome hints at dormant potential. It's there, folded in the quiet corners of our DNA. Not absent. Just silenced.

The cells that escape this silence—like HeLa—demonstrate what happens when the switch flips on again. They divide endlessly. They colonize labs. They replicate across continents, outliving their host by decades. In doing so, they expose a paradox: mortality isn't a fixed feature of the cell. It's an imposed rule.

The question, of course, is why that rule exists. And what happens when we try to rewrite it.

But those questions belong to another part of the story.

For now, we sit with this strange truth: some cells are built to stop. Others are built to go on. And at the center of that divide is a molecule no thicker than a whisper—rebuilding the ends, restarting the count, and offering a glimpse of life beyond the clock.

THE RAZOR'S EDGE

It began, like many things in science, with a question that sounded too simple: if cells age because their telomeres shrink, why not just keep those telomeres long?

The logic was seductive. Activate telomerase. Prevent attrition. Preserve the genome's ends. Arrest aging at its root. Researchers across the world tried it—in petri dishes, in mice, in carefully monitored models. And for a moment, it worked. Old mice with reawakened telomerase showed signs of improvement. Their tissues repaired better. Their stamina increased. Some lived longer than their untreated counterparts. It was, for a while, a whisper of what science had promised for a century: reversal.

But then, another whisper returned: malignancy.

In a separate set of experiments, mice engineered to express telomerase throughout life began to develop tumors. Not all. But many. The very mechanism that had once stopped their cells from aging also allowed some of those cells to ignore every biological checkpoint, to divide with impunity, to refuse death. In trying to keep cells young, researchers had made them unstoppable.

CLOSER LOOK *The two-edged sword of telomerase*

The case for reactivating telomerase looked compelling in 2010, when DePinho's lab at Harvard reactivated telomerase in prematurely aged mice and watched degenerated tissues regenerate — neurons grew back, fertility returned, organ atrophy reversed. It was one of the cleanest demonstrations of biological reversibility in mammalian aging.

The complication is what telomerase also enables. Most cancers are immortal in part because they reactivate telomerase to escape the Hayflick limit; without that step, malignant cells would senesce after enough divisions. Earlier work by Tomás-Loba and Blasco showed that telomerase overexpression in cancer-resistant mice can extend lifespan without driving tumors — but only when the cancer-protective backdrop is in place. Translated to humans, the lesson is a careful one: telomerase is a powerful aging lever, but the same lever opens a cage.

Jaskelioff et al. (2011). Telomerase reactivation reverses tissue degeneration in aged telomerase-deficient mice. Nature, 469(7328), 102–106.

Because that's what telomerase does. It doesn't choose which cells to help. It simply rebuilds the ends of DNA. That means it extends life for both the builders and the saboteurs. A cell on the edge of senescence may regain its function—but a cell already flirting with transformation may gain a passport to multiply uncontrollably.

This is not a design flaw. It is a biological tension baked into our survival. Telomerase is not on or off like a switch. It's regulated with extraordinary precision, turned up in embryos, dialed down in adulthood, deployed briefly in immune crises, and kept largely dormant in tissues where risk outweighs reward. That suppression is not an accident. It's a defense.

Which makes the dilemma personal: What happens if we override it?

Already, clinics in some parts of the world are offering telomerase-activating therapies—marketed as anti-aging interventions, vitality boosters, longevity aids. The data is preliminary, the regulation patchy, the long-term effects unknown. The promise is restoration. The danger is escalation. What if the cells rejuvenated today become the tumors of tomorrow?

The most dangerous molecule in this chapter, it turns out, isn't telomerase. It's certainty.

In science, as in aging, nothing unfolds in a straight line. The telomere story—once billed as the master key to longevity—has grown more nuanced. Mice have long telomeres, but short lifespans. Humans have shorter telomeres, but live far longer. Birds, which often have intense metabolic demands, manage both long lives and considerable telomere wear, thanks to more efficient repair and antioxidant systems.

In humans, the correlation between telomere length and lifespan is real, but complex. Some centenarians have average-length telomeres. Others don't. Lifestyle, inflammation, oxidative stress, and cellular resilience all shape outcomes. Telomeres matter—but they're not the sole meter stick. They're a clock, yes. But the body doesn't listen to just one ticking sound.

KEY TERMS *Vocabulary of the telomere chapter*

Telomere — Repeated TTAGGG sequence at the ends of every chromosome. A buffer zone, not a gene.

Telomerase — The enzyme that rebuilds telomeres. Active in stem cells, germ cells, and most cancers; mostly silent in adult somatic tissue.

Hayflick limit — The number of divisions a normal somatic cell can perform before entering senescence — roughly 50 in human fibroblasts.

Replicative senescence — The non-dividing state a cell enters when its telomeres get critically short.

Dyskeratosis congenita — Inherited disorder of telomere maintenance. Causes bone marrow failure, lung fibrosis, and accelerated aging.

Even within tissues, the effects vary. Muscle, for instance, shows only weak evidence of telomere-driven aging. Its cells divide slowly, if at all. The wear accumulates elsewhere—in joints, in tendons, in neuromuscular junctions. Meanwhile, the immune system, with its relentless cell turnover, shows sharp decline as telomeres run out. One tissue drifts. The other hits a wall.

So telomeres are not a universal script. They are a context-dependent rhythm—silent in some organs, deafening in others.

The siblings from the clinic—Zoya and her brother—had no such buffer. Their inherited disorder stripped them of telomerase function across the board. For them, the clock wasn't ticking slowly in select tissues. It was pounding in every cell. Their marrow collapsed. Their lungs stiffened. Their skin bruised at a touch. They aged not because time passed, but because time had nowhere left to go.

For the rest of us, the story is slower—but not so different. We begin with abundant telomere reserve. Then, year after year, division after division, we draw from that reserve. At first, it feels infinite. Then finite. Then fragile.

By the time most people feel the effects—slower healing, reduced immunity, subtle fatigue—the cellular foundation has already changed. The telomeres are shorter. The responses are slower. The defenses are thinner. It is not visible decay. It is the absence of what once was—a quiet forgetting of youth, written in base pairs.

And in the background, the same question always remains: Could we change this?

Telomerase shows us we might. But it also warns us what that change might cost.

The real lesson isn't that aging is irreversible or that science will one day halt it entirely. It's that biology is less a machine and more a negotiation. Between growth and restraint. Between repair and risk. Between the temptation to intervene—and the wisdom to pause.

The edge of a chromosome is not just a biological detail. It's a boundary. One that separates what we want from what we're built to handle.

And sometimes, to live long is not to extend time—but to respect its limits.

Reader-tool boxes for this chapter

CLOSER LOOK ***What dyskeratosis congenita teaches us***

Dyskeratosis congenita is a rare inherited disorder of telomere maintenance. Mutations in genes such as DKC1, TERC, and TERT cripple the telomerase complex or its accessory proteins, and the result is precisely what you would predict if telomere length were causally upstream of aging: tissues with the highest cell turnover fail first. Bone marrow collapses, lung tissue scars, skin and gut linings thin, and immune systems falter — often before the third decade.

The disease is, in effect, a natural experiment that isolates one hallmark of aging from the others. Patients are not exposed to more oxidative damage or more sun than anyone else; their cells simply run out of telomere reserve too early. That is why a clinic in rural Punjab, looking at two siblings aging at speed, can see the same biology that drives the slow version of the same story in the rest of us.

Bertuch (2016). The molecular genetics of the telomere biology disorders. RNA Biology, 13(8), 696–706.

THE EPIGENETIC DRIFT

ALTERED GENE EXPRESSION OVER TIME

They still shared the same laugh.

Even now, in their seventies, when one cracked a joke about her swollen knees or the other mimicked their childhood piano teacher with impeccable disdain, their laughter harmonized like a single note played in two octaves. Strangers noticed it—at the pharmacy, in the park, in the elevator of the geriatric clinic. "You two must be sisters," people would say, and they'd nod with the knowing glance of women who had been mistaken for each other since birth.

But if the laughter was identical, their faces no longer were.

Maya's cheeks had hollowed, her skin leathery from years in the sun. Her hair, once deep black, was now a wiry gray. A tremor ran through her left hand that she tried to hide, unsuccessfully, by tucking it into the crook of her elbow. Her gait was cautious, almost defensive, as if her bones no longer trusted her muscles. Across from her sat Layla—upright, her eyes alert behind round glasses, silver hair neatly pinned. Her hands were steady. Her voice carried. She still hiked.

They were monozygotic twins—born minutes apart, with genomes so alike you could barely tell one DNA strand from the other. But somewhere along the years, they had drifted.

Maya had stayed in their industrial hometown, married young, raised three children, and worked in a textile mill for most of her life. She smoked for two decades, drank occasionally, and powered through menopause with over-the-counter painkillers. Layla moved to the city, became a schoolteacher, practiced yoga, and ate kale before it was fashionable. They shared a genetic code—but not a lifestyle, not a geography, not a daily rhythm.

Now, their bodies seemed to reflect those differences—not subtly, but starkly.

Scientists used to puzzle over such cases, throwing around vague ideas like "lifestyle," "stress," or "environment" without knowing how to connect those words to molecular facts. How does an argument with your spouse in 1992 leave a mark on your cells? How does the job you hated at forty show up in your skin at seventy? The genome, after all, was supposed to be the map. These women had the same map. Why had they arrived at such different destinations?

The answer, it turned out, lay not in the map—but in how it was read.

There's a myth we like to believe about genes: that they are destiny. That your DNA is a static book, written in ink at conception, and from it your entire life unfolds. But nature, ever more complex than metaphor, doesn't work like that. Your genome is not a script—it's a library. What matters is not what's written, but what's read. And that reading is regulated by a second layer of information: the epigenome.

This layer doesn't change the letters of DNA. It places bookmarks. It highlights passages. It closes entire volumes when they are no longer needed. And over time, this system—elegant, fluid, evolutionarily essential—begins to drift.

This drift is why Maya and Layla aged not just differently, but asymmetrically. It's why Maya's cells began turning on inflammatory genes in her 60s that Layla's kept silenced. Why repair enzymes in Maya's body became sluggish while Layla's still pulsed with youthful rhythm. Their differences were not coded in the DNA—but in how the DNA had been interpreted, chapter by chapter, year after year.

In biological terms, this phenomenon is called epigenetic alteration. In simpler language, it is the body misplacing the conductor's baton. Each cell is an orchestra with access to the full score. But as the decades pass, some instruments go quiet when they shouldn't. Others blare out of turn. The rhythm grows uneven. The symphony turns to noise.

And unlike mutations, which are rare and fixed, these changes are dynamic. They respond to experience. Stress, diet, pollution, exercise, sleep—all the invisible choices and accidents of life leave fingerprints on the epigenome. This is both terrifying and hopeful: terrifying because your biology is not entirely your own; hopeful because change, at least in theory, can be undone.

In the wild, we see epigenetic drift in action—subtle, persistent, and often decisive.

Take the queen bee. She is genetically identical to her sterile sisters, yet she lives ten times longer and reproduces constantly. Her destiny is not written in genes, but in diet.

Fed royal jelly in her larval stage, she undergoes a profound epigenetic transformation that activates fertility genes and silences aging pathways. She becomes, through nurture not nature, a monarch.

Or consider certain fish species like the clownfish. When the dominant female dies, a male can transform into her replacement—not by altering its DNA, but by reshuffling its epigenetic settings. Hormones change. Behaviors shift. Entire body plans are remodeled without a single gene mutation. This is evolution's flexibility at work: not in generations, but in lifetimes.

Humans, of course, don't swap sexes or caste roles. But we do carry that same machinery. And while we can't command it at will, our cells still respond to the long arc of experience. The poverty Layla escaped. The air Maya breathed. The griefs, the triumphs, the routines. They didn't change the DNA. They changed its usage.

This is the first sign that aging is more than wear and tear. It's also miscommunication.

Cells begin to lose their sense of identity—not because their genes are broken, but because they no longer know which ones to use. A skin cell might start expressing immune proteins. A liver cell might forget to turn on detox enzymes. Like a craftsman handed the wrong set of tools, the cell keeps working—but clumsily. The result is dysfunction without disease. Aging, in its earliest form.

Epigenetic drift isn't visible on a scan. It doesn't show up on blood tests. But it's there—quietly redirecting the flow of biology beneath the surface. And it starts long before symptoms appear. In Maya, perhaps, it began with the inflammation from decades of factory dust. In Layla, maybe it was delayed by walks in the park and stress calmed by friendship. The genome was a script. The epigenome was the performance.

As scientists now trace the contours of this second layer, they are beginning to measure it. To model it. Even to modify it. The question is no longer if the epigenome matters—but how much.

And in the story of two aging twins—one bent, one upright—we find the question that defines this chapter: if the genome is the hardware, and the epigenome the software, what happens when the code starts to drift?

Not crash. Not collapse. Just... drift. That's where we begin.

THE SOFTWARE OF LIFE

If DNA is the hardware of life, then the epigenome is its software—an invisible set of instructions that tells each cell which parts of its genetic code to use, when to use them, and how loudly they should be played.

Unlike software on your phone, which you can reinstall or reset with the push of a button, the epigenome is coded in chemistry: molecular marks laid down in layers across your DNA and its scaffolding proteins. These marks don't alter the sequence of your genes, but they profoundly influence how your genes behave. They act like volume knobs, dimmer switches, or sometimes circuit breakers—tuning genes up, down, or off entirely.

In its prime, the epigenome is a miracle of precision.

Picture a fertilized human egg: a single cell carrying the full genetic library for an entire body. As it divides, it produces all the cell types needed for life—neurons, skin, blood, muscle, bone—not by rewriting its DNA, but by reading different chapters at different times. The epigenome is the librarian. It knows that liver cells don't need to read the neuron manual. It ensures that heart cells express cardiac genes but stay silent on the instructions for making hair.

This system allows for extraordinary specialization using the same script. Your muscle cells and brain cells are genetically identical. What makes them different is epigenetic control: chemical signals that determine which genes get activated, which get muted, and which stay in cold storage.

The most well-known of these signals is DNA methylation—the addition of tiny carbonand-hydrogen tags (methyl groups) to the DNA strand. Often, these tags land near a gene's promoter region, acting like red tape to prevent that gene from being copied into action. Then there are histone modifications—chemical tweaks to the protein spools around which DNA is wound. These changes tighten or loosen the packaging, influencing whether a gene is accessible or buried.

When you are young and your body is still developing, these systems run with astonishing coordination. The right genes turn on at the right time in the right tissues. Muscle builds. Organs grow. Skin renews. Memory forms. Every-thing unfolds with clockwork elegance.

But nature, for all its complexity, rarely builds perfect systems. It builds systems that work well enough, long enough, to pass on genes.

And so, over time, the software begins to glitch.

A bit of methylation is added where it shouldn't be. A histone mark is removed too early or placed too late. The packaging of the genome becomes disorganized. Genes that should be silent start to whisper. Genes that should be active fall quiet. The library's lights begin to flicker.

This is epigenetic drift.

It's not caused by a single event. It's the sum of a lifetime of inputs—stress, pollution, diet, trauma, rest, social bonds, and solitude. Each experience leaves its mark on the epigenome. Some changes are adaptive. Others are random. But over the decades, the patterns lose coherence. The genome doesn't break. It just forgets how to sing.

In nature, you can see the same pattern emerging—when biological software loses its way.

Among wild Pacific salmon, for instance, there is a tragic clock embedded in their biology. After years at sea, they return to the rivers where they were born, spawn once, and die. This terminal program is not triggered by injury or infection. It's encoded epigenetically: a cascade of gene silencing and activation set off by environmental cues like water temperature and daylight. As soon as spawning begins, immune genes go quiet, muscle tissue begins to waste, and metabolic control unravels. The salmon's body is rewriting itself toward death. It's not a failure. It's a feature. Epigenetics controls the exit.

In plants, too, epigenetic control can dictate life and death. A biennial plant like the foxglove lives for two years. During its first year, it focuses on growing roots and leaves. Only after a cold winter—signaled through changes in its epigenetic marks—does it flower and seed in the spring. If you disrupt those marks artificially, the plant forgets to flower. The clock stalls. It cannot reproduce. Epigenetics is not just about when things begin. It's about when they're allowed to end.

Evolution doesn't waste energy preserving perfect function forever. It calibrates for the survival of the next generation. The epigenome follows suit. Early in life, it's heavily policed—corrected, pruned, updated. Later, those corrections slow. Surveillance weakens. The software starts saving in the wrong file formats.

In humans, this drift begins quietly. A skin cell might slightly underexpress its DNA repair genes. A lung cell might fail to switch off a proinflammatory gene. At first, these deviations are minor. But as they accumulate, they reshape how the body functions—not because the genes are wrong, but because the instructions are garbled.

This, more than mutation, explains much of aging's slow creep.

In aged muscle, genes that support regeneration go quiet, while genes that promote fibrosis—the buildup of scar-like tissue—become louder. In aging brains, neurons lose the methylation marks that normally silence immune genes, leading to low-grade inflammation even in the absence of infection. In the liver, detoxification genes may fall silent. All of it is epigenetically controlled. All of it part of the drift.

The most haunting aspect is that much of this misregulation is reversible—at least in theory.

In lab experiments, scientists have taken old cells and reset their epigenetic marks, restoring youthful gene expression without altering the DNA itself. This doesn't fix mutations. But it revives the cellular "mood." It's as if the conductor picks up the baton again, and the music, once dissonant, finds its rhythm.

This idea—that the information of youth still exists, hidden beneath the drift—suggests something staggering: aging may not just be about damage. It may be about memory loss. Not yours. Your cells'.

Of course, evolution never promised longevity. It promised reproduction. And so, the system that gave us youth was never designed to last. But now, with tools that can read, map, and even modify the epigenome, we are beginning to ask: what if it could?

Before we can answer that, though, we must first understand how far the software has strayed. We must read the corrupted code.

That's where we turn next: to the scientists who mapped this drift, one methyl group at a time—and discovered that the history of your life is written not just in your wrinkles or your aches, but in the silent marks left on your DNA.

MAPPING THE CLOCK

It began, as many breakthroughs do, with a hunch and a spreadsheet.

Steve Horvath wasn't looking to redefine the way we measure age. At the time, he was a biostatistician at UCLA, working in the strange borderlands between data science and molecular biology. His job, in essence, was to find patterns in the chaos—to tame the staggering rivers of genomic data being generated by modern labs and see what stories they might be trying to tell.

But age kept getting in the way.

No matter what tissue he analyzed—liver, blood, brain, saliva—he noticed the same thing: a set of molecular signatures that appeared to change with the number of birthdays behind a sample. They weren't mutations. They weren't signs of wear and tear. They were chemical tags—methyl groups—placed delicately on cytosine bases across the DNA. The pattern wasn't random. It was eerily consistent.

It was as if the genome was leaving timestamped fingerprints on itself.

Most people assume that age is something you see—on a face, in a posture, across the surface of the skin. But what Horvath began to suspect was that age could be read—literally decoded—from inside the cell. Not from the DNA sequence, which stays mostly the same throughout life, but from the epigenetic marks that frame it.

So he began to test the idea.

Using public datasets—dozens of them—he compiled methylation profiles from thousands of tissue samples. These were anonymized files meant for other kinds of studies. But to Horvath, they held a hidden message. He filtered out the noise, cross-referenced sites across tissues, and looked for methylation changes that correlated most tightly with chronological age.

And then, with a few lines of R code and a moment of disbelief, he built a model that could predict a person's age—within a few years—based solely on the epigenetic state of their DNA.

He called it an epigenetic clock.

It wasn't magic. It was statistics. But it felt like sorcery. You could take a blood sample from a stranger, run it through this algorithm, and learn not how long they'd been alive—but how long their cells believed they'd been alive.

For some people, the number matched their birthday. For others, it didn't. Some 40-yearolds had the epigenetic profile of someone ten years older. Others looked biologically younger than their license said. And that gap—the difference between chronological and epigenetic age—began to matter.

People with olderlooking methylation profiles were more likely to suffer from chronic disease. They were more likely to die sooner. They aged faster—not in the mirror, but in their biology. Horvath's clock had revealed a second timeline, running quietly beneath the visible one.

It was like discovering that everyone carries two pass-

ports: one stamped by time, the other by biology. And sometimes, they tell different stories.

The implications rippled quickly.

If aging could be measured at the molecular level, it could also be tracked, studied, and maybe—one day—slowed. Clinical researchers began testing the clock against interventions: Would exercise affect the tick rate? Meditation? Diet? Chemotherapy? Could a stressful life leave epigenetic scars that made the clock run faster? Could resilience make it tick more slowly?

Some answers came fast. Smoking, for instance, aged the epigenome by several years. Childhood trauma left a methylation shadow decades later. Poor sleep, poor diet, loneliness—all left fingerprints on the clock. But so did protective factors. Regular movement. Social connection. Curiosity. They didn't stop time. But they slowed the ink.

In this way, the epigenetic clock became something more than a research tool. It became a mirror, held up to the choices and exposures that make us who we are.

Nature, as always, had foreshadowed this.

In wild animals, survival often hinges not on raw genetics but on adaptability—on the ability to turn certain genes on or off in response to environment. Take the Arctic ground squirrel, which survives brutal Alaskan winters by entering a state of suspended animation. As its body temperature drops near freezing, its epigenetic marks change dramatically—shutting down genes for growth, metabolism, even cognition. When spring returns, so do the genes. It wakes not just from sleep, but from silence. An epigenetic time machine.

Or the naked mole rat, that subterranean mammal with negligible signs of aging and almost no cancer. Its epigenetic drift is incredibly slow. Methylation patterns in its tissues remain stable for decades. The body forgets how to die, simply because it never stops remembering how to function.

In contrast, domesticated animals like lab mice—bred for speed, not longevity—show rapid epigenetic drift. Their clocks tick fast. Their lives are brief. Biology, it seems, does not reward haste with endurance.

Horvath's clock also revealed something humbling. Some tissues age faster than others. Blood and liver show rapid epigenetic changes. The brain, more slowly. Certain parts of the hypothalamus—the brain's command center for hormone regulation—barely budge, epigenetically, even into old age. It's as if time passes at different speeds inside different rooms of the same house.

Even more uncanny: transplanted organs seem to "adopt" the age of their new host. A young liver placed in an older person begins to show epigenetic signatures of aging. It adjusts. It listens. The clock is not an isolated machine—it's a system, embedded in the body's broader symphony.

In the end, Horvath's insight wasn't just about measurement. It was about understanding aging not as a slow failure, but as a process of coordination gone awry. The methylation patterns that once turned the genome into a harmonious ensemble begin to drift out of tune. The conductor is still there. But the baton wavers.

And yet, the notes remain.

That's the strange, hopeful truth embedded in the clock: the sheet music is still legible. Even as cells age, the capacity to remember youth—to re-access a younger epigenetic state—remains. The drift is reversible. At least in theory.

So when Horvath stared at his charts, what he saw wasn't just decline. He saw a map. A path back. A way to quantify resilience, recovery, and possibly—even rejuvenation.

The challenge, now, is to go from reading the clock to resetting it. To take that silent rhythm of methyl groups and learn how to rewrite it—without breaking the music.

And for that, we must turn to the next act of this story: the scientists who dared to rewind.

REWRITING THE MUSIC

Every symphony has a score. And every score, at some point, starts to go off-key.

In the aging body, that score is the epigenome—the delicate sheet of chemical marks that instructs each cell what to do, when to divide, how to behave. As time marches on, these marks blur. The conductor misplaces pages. Violins come in late. The brass section drones off beat. The music of youth, once fluid and precise, begins to sound like static.

For years, scientists believed this score was permanent. That once a cell had committed to its role—muscle, neuron, liver—

it could never return to an earlier state. That the music, once played, could not be unplayed.

Then came a biologist named Shinya Yamanaka, who asked a question no one else had dared.

What if the score wasn't fixed? What if it could be reset?

It was 2006, and Yamanaka's lab in Kyoto was trying something radical: taking fully mature cells—skin cells, mouse fibroblasts—and exposing them to a cocktail of just four transcription factors: Oct4, Sox2, Klf4, and c-Myc. These weren't exotic ingredients. They were common regulators, found in abundance during embryonic development. But when combined in just the right way, they did something extraordinary.

They made old cells forget who they were.

Not in a destructive way, like wiping a hard drive. But in a precise, uncanny reversal. The cells lost their adult identity. They shed their methylation marks. They returned to a pluripotent state—capable of becoming anything again. Bone. Brain. Heart. Skin. These were no longer skin cells. They were induced pluripotent stem cells, or iPSCs: biological blank slates.

It was as if someone had taken a smudged, weathered manuscript and rewritten it in fresh ink—restoring the original melody with perfect clarity.

The implications were staggering. For the first time, scientists had found a way to push the biological clock backward—not in a metaphorical sense, but literally. The epigenetic markers of age, the molecular graffiti scrawled across the genome, could be scrubbed clean.

And yet, like all discoveries that dare to rewrite nature, it came with danger.

Wipe too much of the score, and the orchestra forgets not just its timing—but its identity. A skin cell that forgets it's a skin cell can become anything. Including, sometimes, a tumor.

Yamanaka's breakthrough earned him the Nobel Prize, but it also forced a new caution: if reprogramming could reset age, it could also erase order. The question became not whether we could rewind the clock—but how far, how fast, and how safely.

In evolution, identity is everything. A coral polyp must know how to build reef. A hummingbird cell must know how to metabolize sugar at 1,200 beats per minute. That identity is encoded not just in DNA, but in the layers of instruction wrapped around it. Methylation marks. Histone tags. Chromatin structure. These layers are honed over millions of years—not for flexibility, but fidelity.

Yet here was evidence that identity wasn't as rigid as we thought. That age could be undone. That cells could be coaxed back into youth.

The natural world has always flirted with such possibilities. Take the axolotl, that strange Mexican salamander with a permanent smile and the power to regenerate entire limbs. When injured, its cells near the wound site dedifferentiate—stepping backward in development—before reorganizing into muscle, bone, and skin. The axolotl doesn't just heal. It replays the music of growth, precisely and cleanly.

Or the jellyfish Turritopsis dohrnii, sometimes dubbed the "immortal jellyfish." In times of stress, it reverses its life cycle, transforming from mature medusa back into its polyp stage. It doesn't age—it loops. Reprogramming, in nature, isn't just possible. It's ancient.

What Yamanaka's factors offered was a way to bring that logic into mammals—into us.

In the years that followed his discovery, labs across the globe raced to adapt the technique. They refined the dosage. They shortened the exposure. They asked: Could we use reprogramming not to erase identity, but to restore function? A team at Harvard, in collaboration with David Sinclair's lab, tried it on old mice. Not the whole body—just the retina. The eyes of these mice had been damaged by age. Vision dulled. Cells stiffened. The marks of epigenetic drift—methylation noise, silenced repair genes—were everywhere.

They applied a partial reprogramming protocol. A gentle rewind. Enough to polish the CD, not smash it.

And the results were electric.

Vision returned. Not perfectly, not universally—but measurably. The epigenetic marks shifted toward a youthful pattern. The cells began to function like they had years before. These were not embryonic cells. They were not stem cells. They were rejuvenated retinal neurons.

The experiment was published in Nature. The headlines followed. "Biological Age Reversed in Mice." "The Future of Aging?" But the real revelation lay deeper: the DNA remained unchanged. The damage wasn't fixed. The mutations were still there. What had changed was the software.

The music was clearer.

This ushered in a new paradigm: perhaps aging wasn't the result of broken hardware, but corrupted instructions. Not a scratched record, but a scrambled signal. And if that signal could be reset—if the cells could remember how to perform again—then aging itself might be, at least in part, a problem of memory.

Of course, the risks remain enormous.

Turn back the clock too far, and cells may lose track of their place in the body's ecosystem. A reprogrammed liver cell that forgets it's in the liver is a threat, not a remedy. The balance between youth and identity is razorthin. And in a system as complex as the human body, reintroducing pluripotency is like opening the gates of a dam—caution is the only safeguard.

But the horizon has shifted.

We now live in a world where the idea of cellular rejuvenation is not science fiction. It's a lab protocol. Where the notion that a neuron might recover lost function—or a muscle cell regain youthful resilience—has moved from the improbable to the plausible.

And in that new landscape, the epigenome is no longer a mystery. It's a manuscript. One we are slowly learning to edit.

Not to rewrite who we are. But to help us remember who we once were.

DANGEROUS MEMORY

There's a story told among beekeepers: if you move a hive too far from its original spot—say more than a few feet—the bees become disoriented. They spiral back to where home used to be, circling the air, searching for a memory that no longer matches reality. The coordinates in their mind are fixed. But the landscape has changed.

Cells, it turns out, aren't so different.

Within every nucleus is not only DNA, but the memory of how that DNA has been used. A skin cell remembers how to be skin not just through its genes, but through the orchestration of its epigenome—what's turned on, what's silent, what's poised. It's this epigenetic memory that gives order to the body. It's what makes 200 types of cells all perform different tasks with the exact same genome.

But memory, powerful as it is, can become dangerous. That danger comes into focus when we try to reprogram it.

After the early success of Yamanaka's factors in resetting age, a question lingered in every lab brave enough to try it: if you tell a cell to forget its age, what else might it forget?

The answer, as it turned out, was both thrilling and sobering. Push reprogramming too far, and the cell loses its identity entirely. A heart cell might lose its contractile rhythm. A neuron might abandon its axonal shape. The cell doesn't die. It just becomes something else. Or worse—something unstable. Something with no clear purpose.

Something dangerous.

In 2014, researchers experimenting with partial reprogramming noticed a sharp boundary between rejuvenation and chaos. In short spurts, the Yamanaka factors improved cellular function. But longer exposures triggered a loss of cell type. The tidy signature of muscle became smeared. The immune cells forgot how to recognize invaders. The tissue landscape became a blur. The cells were alive—but rudderless.

In that moment, reprogramming revealed its paradox: it could restore youth, yes—but it could also erase the very thing that made a cell safe.

Cancer, in many ways, is the outcome of such erasure. Tumor cells are not inherently foreign. They are us—misremembered. Cells that, in the scramble of replication, accumulate mutations in oncogenes, yes—but also shed the epigenetic marks that once tethered them to order. A pancreatic cell that once turned off division-related genes now keeps them active. A colon cell that once respected contact inhibition now ignores it. The instructions are there. The brakes are gone.

And telomerase, the molecule so promising in anti-aging trials, is right there helping them.

It's no coincidence that the vast majority of human cancers show reactivation of telomerase. Once a cell has begun to forget its identity, telomerase ensures it has the time to finish the transformation. Divide, survive, mutate, repeat. A rogue factory with no shutdown switch.

This is why the push toward anti-aging therapies based on reprogramming is moving slowly—perhaps wisely. Because in our rush to rewind, we risk removing the guardrails that evolution spent hundreds of millions of years building.

Evolution, after all, didn't optimize us for eternal youth. It optimized us for survival to reproduction—and a little beyond. Longevity, in that framework, is a luxury. Not a priority. The epigenetic system we inherit is built not for permanence, but adaptability—changing gene expression in response to environment, nutrition, stress. That very plasticity is why we survive. But it's also why things can go wrong.

Epigenetic drift—the random shifts in gene regulation that come with age—isn't just decay. It's the price of flexibility. The system was never meant to hold one pattern forever. It was meant to respond. But over time, the signalto-noise ratio worsens. The response becomes erratic. A little inflammation turns into chronic immune activation. A single round of stress leaves permanent scars. The flexibility becomes fragility.

This too can be seen in nature.

Consider the octopus, a creature of staggering intelligence and grace, capable of solving puzzles, escaping jars, even recognizing individual humans. Yet the octopus lives just a few years. Its lifespan is tightly governed by its reproductive cycle. Once it mates, a cascade of gene expression changes begin. The mother stops eating. Her brain rewires. She guards her eggs, then dies.

What triggers this death program?

Not genetic mutations. Not damage in the traditional sense. But a change in gene expression. A rewriting of the instructions. In essence, an epigenetic shift that says: your role is complete.

If we tried to "fix" that—if we halted the epigenetic cascade, kept the octopus alive after mating—what might remain? An organism disconnected from its natural script? A mother without a role? The science isn't there yet, but the metaphor is strong: sometimes, restoring youth might mean severing meaning.

That's the razor's edge scientists now walk.

Partial reprogramming holds undeniable promise. In mouse models, intermittent exposure to Yamanaka factors has restored organ function, reversed signs of DNA damage, and even improved cognitive markers. But the protocols must be exact. A little too much, and tissues collapse into disarray. A little too long, and identity is lost.

So the field now turns to finetuning.

New tools, like CRISPRguided epigenetic editors, allow scientists to change specific methylation marks without touching the DNA sequence. It's like erasing just one wrong note in the orchestra without disturbing the whole score. Others are exploring chemical compounds—diet-derived or pharmacological—that nudge the epigenome gently, coaxing old cells to behave a bit younger without making them forget what they are.

Even the concept of "epigenetic programming windows" is emerging: times in life—perhaps early childhood, or post-injury—when the epigenome is especially malleable. Interventions in these windows may offer high reward with lower risk.

But the core dilemma remains: how do you make a cell remember youth without making it forget itself?

That question is not just scientific. It's philosophical.

Because in the end, memory is the thread that holds biology together. Memory of damage. Memory of development. Memory of what it means to be part of a tissue, an organ, an organism. Rewriting that memory is powerful. But it is also perilous.

Just ask the bees. The Song Beneath the Static

Somewhere in the forests of New Zealand, there lives a bird that forgets how to sing.

The hihi, or stitchbird, once sang elaborate, regionally distinct songs passed from parent to fledgling like a family heirloom. But in certain populations—particularly those reintroduced into new habitats by conservationists—young males have begun to sing oddly. Their songs lack structure. They stumble on phrases. They improvise or fall silent altogether. Not because they are physically incapable of singing, but because no one taught them the melody. The chain of transmission was broken.

The loss isn't genetic. The birds still carry the DNA of song. The loss is cultural, environmental, epigenetic in spirit. The hardware is intact. The software is scrambled. A system once synchronized by lineage and learning has fallen into noise.

Aging, some scientists now argue, might be a similar kind of forgetting.

For decades, biology approached aging as a matter of damage—accumulated mutations, oxidative stress, cellular wear. But a quieter theory has emerged: perhaps aging is not just the presence of damage, but the loss of coordination in how cells read the script of life.

This is the theory behind epigenetic drift. Over time, the systems that once told each cell which genes to express, when to divide, when to rest, begin to go offbeat. A conductor with a shaky hand. A signal smudged with interference. The cell still holds the blueprint, but it plays the wrong part, at the wrong time, or too softly to matter.

It's as if the song of youth is still written in the score, but static has overtaken the speakers.

This theory would have remained speculative if not for what happened in 2023. In a landmark experiment, researchers took healthy adult mice and, without introducing any DNA mutations, deliberately scrambled their epigenetic marks. Not randomly, but systematically—methylation here, histone shifts there—nudging the system out of alignment. The result? The mice aged rapidly. Their tissues degraded, their cognitive function declined, their physiology mirrored that of older counterparts.

Then came the reversal.

The same researchers applied a partial epigenetic reset using tightly controlled reprogramming factors. They didn't fix the DNA. They didn't remove any damage. All they did was restore the epigenetic settings closer to their youthful state.

The mice recovered.

Vision improved. Tissue integrity returned. Certain organs functioned as if the passage of time had been edited backward—not erased, but softened. It was like clearing the distortion from a scratched record, letting the original melody shine through.

To some, it was the most compelling evidence yet that aging is, at least in part, a loss of information flow. Not just the degradation of parts, but the misreading of instructions.

The implications are enormous.

If aging is a drift—a static layered on top of biological coordination—then the path forward might not lie in gene editing, or damage repair, or calorie restriction alone. It might lie in restoring pattern. In reminding cells how to sing in harmony again.

This brings us to a concept borrowed from another field: systems theory.

In systems biology, function arises not just from parts, but from relationships. Feedback loops. Rhythmic cycles. Emergent behavior. In youth, these systems are coherent. The liver responds to feeding. The brain cues rest. The gut, the skin, the immune system—all dance to the same tempo.

With age, the coherence weakens. The rhythm slips. Circadian genes, once tightly linked to sunrise and feeding, begin to drift. Hormonal pulses lose amplitude. Inflammatory responses become overactive or blunt. It's not that any one system is broken—it's that the orchestra no longer plays in time.

The same can be seen in ecosystems.

Coral reefs, for instance, can maintain stunning biodiversity as long as balance is preserved—light, temperature, nutrient flow. But a subtle shift in any parameter—water too warm, light too dim, algae too abundant—and the whole reef changes character. The coral bleaches. Fish migrate. Predators arrive. No single gene changes. No single animal mutates. The system simply tips.

And the recovery, if it happens at all, requires more than fixing a single species. It requires restoring the web of feedback. The synchrony.

So too, perhaps, with aging.

What if the goal isn't to make a cell immortal, but to restore its sense of timing? What if the cure for aging is not found in erasing age, but in restoring rhythm?

The idea is still young. But research is pointing in that direction.

Already, scientists are exploring compounds that influence epigenetic readers—enzymes that interpret methylation marks and histone codes. Others are studying how deep sleep, fasting, or exercise can recalibrate circadian and epigenetic programs. The interventions don't extend telomeres. They don't edit genes. They modulate pattern. They tune the song.

And the most radical implication?

That the instructions for youth may never leave. They may not be lost. Only buried.

That beneath the errant marks, the offbeat rhythms, the noise of years—there may still be, in every cell, the original sheet music. Waiting to be read again.

Which brings us back to the twins.

Decades after their lives diverged, epigenetic tests showed that the sister who smoked, endured hardship, and bore chronic stress had an epigenetic age nearly ten years older than her biological twin. The DNA was the same. But the software—the orchestration—was different. She hadn't just aged faster. She had drifted farther.

And yet, even in her cells, the original code remained.

Not pristine. But present.

This, in the end, is the hope that epigenetics offers. That aging, while inevitable, might be influenceable.

That time's signature isn't only carved into our skin or bones—but written in a language we're finally learning to read.

And perhaps—just perhaps—to rewrite.

Reader-tool boxes for this chapter

CLOSER LOOK ***Horvath's clock — a calendar written in methyl groups***

In 2013, UCLA biostatistician Steve Horvath published a paper that did something the field had been trying to do for decades: he built a biological age estimator that worked across nearly every human tissue. By analyzing methylation levels at 353 specific cytosine-guanine sites, he could predict a tissue's age within a few years.

The clock was not just descriptive. Cancer tissues looked decades older than their host. Newborn samples read close to zero. Induced pluripotent stem cells reset to nearly zero — the act of reprogramming a cell back to an embryonic state, it turned out, also rewound its epigenetic age. The implication was startling: the methylation clock isn't an incidental marker of time passing. It moves with the biology, and it can move backwards.

Horvath (2013). DNA methylation age of human tissues and cell types. Genome Biology, 14(10), R115.

EASY ACCESS ***Genes vs. epigenes — what's the difference***

Your genome is the hardware. Three billion letters of DNA, mostly the same in every cell of your body, mostly the same as the day you were born. Your epigenome is the software — chemical marks layered on top of DNA that decide which genes get read out and which stay silent. A liver cell and a neuron carry the same hardware; their epigenomes are wildly different. That's why one makes albumin and the other fires action potentials. Aging, increasingly, looks like a slow corruption of the software, not the hardware.

CLOSER LOOK *Yamanaka factors and partial reprogramming*

In 2006, Shinya Yamanaka's lab discovered that just four transcription factors — Oct4, Sox2, Klf4, and c-Myc, now called OSKM — could revert a fully differentiated adult cell back to a pluripotent stem cell state. The discovery won the Nobel Prize in 2012 and changed regenerative medicine.

The aging field's question was different: what if you used those factors briefly, partially, in living animals — not to turn cells back into stem cells but to nudge their epigenetic age younger without losing identity? In 2016, the Belmonte lab used cyclic, short OSKM expression in progeroid mice and saw delayed aging without tumor formation. In 2020, Lu et al. showed that a three-factor combination (OSK, leaving out the cancer-prone c-Myc) restored youthful methylation patterns in retinal neurons of glaucomatous mice and restored vision. Partial reprogramming has since become one of the most-watched bets in geroscience — and one of the most carefully hedged ones.

Lu et al. (2020). Reprogramming to recover youthful epigenetic information and restore vision. Nature, 588(7836), 124–129.

CLOSER LOOK ***The information theory of aging***

David Sinclair's 2023 ICE-mouse experiments tested a stark hypothesis: that the act of repeatedly repairing DNA double-strand breaks — a normal, daily occurrence — slowly degrades the epigenetic landscape, even when the underlying DNA sequence is restored perfectly. In the ICE model, mice received non-mutagenic DNA cuts. Their genomes stayed intact. Their epigenomes drifted. They aged.

Then the team applied OSK reprogramming. The aging signs partially reversed. The takeaway, if it holds up to replication, is that aging may be less about damage to the genetic blueprint and more about loss of access to it — like a library where the books are intact but the catalog has been scrambled. The information is still there. The system has just forgotten how to find it.

Yang et al. (2023). Loss of epigenetic information as a cause of mammalian aging. Cell, 186(2), 305–326.

KEY TERMS *Vocabulary of the epigenetic chapter*

Epigenome — Chemical marks on DNA and its packaging proteins that decide which genes are active in a given cell.

DNA methylation — Addition of a methyl group to a cytosine base. Generally silences gene expression.

Methylation clock — A statistical model that estimates biological age from methylation patterns at specific sites. The Horvath clock uses 353.

Yamanaka factors (OSKM) — Four genes — Oct4, Sox2, Klf4, c-Myc — that together can rewind an adult cell to a pluripotent state.

Partial reprogramming — Transient OSK or OSKM expression aimed at rejuvenating an adult cell without losing its identity.

HOUSEKEEPING BREAKDOWN

LOSS OF PROTEOSTASIS

In July 1968, New York City experienced a garbage strike. For nine days, sanitation workers refused to collect trash. Mountains of black plastic bags I ballooned on sidewalks. The city reeked. Rats multiplied. Fires broke out in empty lots filled with rotting refuse. Tourists gagged. New Yorkers—resilient as ever—tried to pretend it was just part of city life. But by day four, no one could deny it: civilization, at least in its polished form, was rotting at the edges.

The city, it turned out, had far more in common with a living cell than anyone realized.

Cities are systems. So are cells. Both produce, consume, and—critically—clean. We talk about growth and energy when we speak of cities, of GDP and expansion and movement. Rarely do we speak about what happens when the basic, invisible process of cleanup breaks down. Yet garbage removal, in biology as in urban life, is not a background detail. It is the central, invisible act that allows everything else to flourish.

Cells, like cities, have waste. And when they stop taking out the trash, they age.

At the heart of this biological housecleaning is a concept called proteostasis—a portmanteau of "protein" and "homeostasis." Proteostasis is not just about getting rid of bad proteins. It's about balance. Maintenance. Continuity. Every second, our cells are building proteins, folding them into precise three-dimensional shapes, sending them to do their work, then degrading them when they are misfolded, damaged, or simply obsolete. This is not a once-in-a-while spring cleaning. This is minuteby-minute sanitation.

And it is not optional.

In youth, this machinery hums. Proteins are born, guided by chaperone molecules that fold them like origami masters shaping paper cranes. The misfolded are tagged discreetly with a molecular kiss of death—a small protein called ubiquitin—and shuttled to the proteasome, the cell's trash compactor. Larger debris or dysfunctional components are swept up by autophagy, a process that literally means "self-eating." Autophagy wraps waste in vesicles and digests it, recycling nutrients and clearing space.

In a young body, this all works so well that we never notice it.

But aging begins when this process falters. The trucks stop running.

Alain is a Frenchtrained chef in his mid-sixties who runs a beloved bistro in Montreal. For decades, Alain has lived by precision. Knife cuts, sauce reductions, the perfect consistency of pâte à choux. His hands have always known what to do before he tells them. But lately, something strange has started happening. At first it was a tremor. A missed grip. A dropped whisk. Then, slowly, words began slipping through his fingers too—names of ingredients, the order of a dish.

He is too proud to say anything. But his apprentice, Noémie, notices.

She says nothing at first, just begins doing more prep. But one day Alain forgets the temperature of his own oven. The same oven he's used every day for the past twenty years.

A week later, he sees a neurologist. Then an MRI. Then the whisper of a diagnosis: earlystage Parkinson's disease.

The doctor explains what's happening in his neurons. There's a protein—alpha-synuclein—that's misfolding. It's sticking to itself, forming aggregates. These clumps are toxic. They're gumming up the works in his brain's basal ganglia, interfering with dopamine production, disrupting the control of movement. And the problem isn't the DNA—his genes are fine. The problem is cleanup. Proteostasis has started to fail.

In biological terms, this is what happens when garbage collection stops.

Proteins are functional only when folded correctly. The process of building them from scratch—translating a gene into a string of amino acids—is only the beginning. What follows is an intricate dance of folding, coiling, tucking, and stabilizing, often assisted by chaperone molecules like HSP70 and HSP90. These heatshock proteins, discovered in the 1960s when fruit flies were exposed to high temperatures in a lab accident, are the cell's first responders to stress. They help salvage proteins at risk of misfolding.

In youth, the cell makes plenty of these chaperones. But as we age, their expression wanes. And the protein load increases. Errors accumulate. The folding game becomes harder.

Misfolded proteins are dangerous not because they're broken, but because they're sticky. Like halfchewed gum, they latch onto other proteins, forming aggregates that clog the machinery of the cell. In neurons, this is catastrophic. But it doesn't stop there. Misfolded proteins also build up in muscle tissue, in the eye lens, in the liver, even in immune cells.

And crucially, in most of these cases, the DNA is still intact. The cell's blueprint is unbroken. But the mess of unfolded, mismanaged proteins overwhelms the system.

Just like New York during the garbage strike, the problem isn't infrastructure. It's management.

In Alain's case, the diagnosis means rethinking every-thing. The neurologist refers him to a specialist at a biotech clinic exploring neurodegenerative diseases. There, a post-doctoral researcher explains to Alain's daughter how proteins, when misfolded and not properly cleared, begin to "aggregate" like uncollected garbage bags. Her words are careful. Her tone is gentle. "It's not that his cells are dying because of DNA damage," she says. "They're dying because the proteins inside them are getting\...clogged. Like a kitchen drain."

It's a metaphor the daughter understands.

She continues, explaining that the cell's disposal units—proteasomes and autophagy—are declining with age. "The system is still there," she says. "But it's not fast enough anymore. It's like trying to run a restaurant with one dishwasher."

Scientists have known since the early 2000s that proteostasis collapse is a defining feature of aging. In experiments with C. elegans, Drosophila, and mice, researchers found that artificially boosting chaperones or proteasome function can extend lifespan. Meanwhile, aged tissues—especially brains—are riddled with protein aggregates. In Alzheimer's disease, these take the form of amyloidbeta plaques and tau tangles. In cataracts, crystallins clump in the eye lens. Even cartilage loses resilience due to cross-linked proteins that should have been cleared.

This is not just neurodegeneration. It is whole-body housekeeping failure.

Cells age not because they lack instructions, but because they lose the capacity to keep their space clean.

We often talk about aging as the accumulation of damage. But more precisely, aging is the accumulation of unresolved mess. A good cell, like a good kitchen, can tolerate a spill. But without cleanup, those spills become crusted pans, stuck utensils, rotting smells.

THE FOLDING GAME

Long before humans invented protein powders or labgrown meat, nature had already perfected the art of making structure from strings. The silk of an orbweaving spider, for instance, is made of nothing more than proteins—and yet, weight for weight, it's stronger than steel. A spider doesn't read blueprints to make silk. It doesn't use fire or forge. Instead, it spins a string of amino acids from its body, folds it with precision, and ejects it into the air, where it hardens into a flawless thread. This is proteostasis at its most elegant—folding, shaping, deploying.

Proteins, after all, are not just molecules. They are machines—dynamic, moving, functional. Each one must be folded just right to do its job. A protein's shape is its identity. Misfolded, it is useless at best and toxic at worst.

To appreciate aging, you have to understand folding. Inside every living cell, protein folding is an extraordinarily delicate performance. When a new protein is created—a linear chain of amino acids fresh from the ribosome—it's like a piece of raw dough. The cell doesn't just throw it into the oven and hope for the best. Instead, it guides the dough, kneads it, folds it—using chaperone proteins—into something recognizable. Imagine a baker trying to shape croissants blindfolded, relying only on touch. That's what cells do millions of times per day.

The correct fold allows a protein to bind, move, catalyze, signal. A misfolded one is like a wrench dropped into an engine—at best, it just jams things up. At worst, it triggers chain reactions of dysfunction. It's about flooding. The cell's equivalent of the folding table is overwhelmed. Debris accumulates faster than it can be cleared.

Back in the 1980s, a Japanese scientist named Kazutoshi Mori was studying yeast cells under stress. He noticed something odd: when the endoplasmic reticulum (ER)—the part of the cell where many proteins are folded—got overloaded, it fired off a kind of molecular panic signal. This became known as the unfolded protein response (UPR), and it's a critical part of how cells sense and react to folding failure.

The UPR is like a smoke alarm in a bakery. Too many burnt croissants? Something's wrong with the oven. It halts production, recruits help, and tries to restore order.

But the UPR can only do so much. If the stress continues

—if misfolded proteins keep piling up—it can initiate cell suicide. Better to destroy one cell than let it spread the chaos.

This response is ancient. It's been preserved across species, from yeast to humans. That's how important protein folding is. It's not optional. It's existential.

Let's zoom out. Evolution has never been about perfection. It's about

tradeoffs. And one of evolution's cleverest tradeoffs is the use of proteins. They're complex, powerful, and versatile—but they're also fragile. They can unfold with heat, with acid, with oxidation. Nature made us powerful engines—bodies that move, think, fight infection, digest—but at the cost of constant maintenance.

Aging, then, is not the accumulation of damage so much as the accumulation of unfixed damage. And in many ways, protein misfolding is the original sin.

There's something poetic here.

Imagine a world-class orchestra slowly losing its sheet music. One violinist starts to play off-key. The cellist follows. The conductor waves his arms faster, hoping to realign them. But the damage spreads. Over time, the entire symphony sounds slightly... wrong. Not broken. Not silent. But out of tune.

That's what happens to the proteome in aging cells. The music goes awry.

And yet, there's hope. Because some of these processes are reversible. We've seen in animal models that mild stress—heat, fasting, certain plant compounds—can reinvigorate chaperone production and autophagy. Like sending a burst of funding to a struggling sanitation department, these interventions buy time.

WHEN CLEAN-UP FAILS

In 1969, an unusually hot summer swept through Denmark. A lab technician working with fruit flies noticed something odd: her specimens, normally active and fertile, were suddenly sluggish, misshapen, sterile. But it wasn't genetic. When temperatures cooled, the flies recovered. The anomaly, initially dismissed as lab error, sparked an investigation that uncovered something profound—heat shock proteins. These were molecular helpers, activated by stress, whose job was to stabilize and refold other proteins under duress.

It was the cellular equivalent of firefighters rushing into a burning building—not to put out the flames, but to save what could still be salvaged.

This discovery changed everything. It showed that cells have contingency plans. But it also hinted at a terrifying reality: when the contingency fails, collapse follows fast.

Let's return to our core theme—proteostasis. Protein homeostasis isn't a single process; it's a choreography. You have production (synthesis of new proteins), folding (shaped by chaperones), quality control (inspection checkpoints), and waste management (via the proteasome and autophagy). Aging is what happens when every step in that dance starts missing beats.

The proteasome—the cell's garbage disposal—begins losing efficiency. This cylindrical structure, shaped like a barrel, normally pulls in misfolded or damaged proteins tagged with a molecular "kiss of death" known as ubiquitin. Once inside, these proteins are shredded into harmless peptides. But by middle age, studies in human cells show that proteasome function drops significantly. Damaged proteins linger longer. Worse, the system clogs. Accumulated junk begins interfering with the machinery itself, slowing it further—a vicious cycle.

Autophagy, the other major cleanup mechanism, falters in parallel. This process, which literally means "self-eating," is essential not only for clearing misfolded proteins, but for removing damaged organelles, pathogens, even entire parts of the cell. It's the Marie Kondo of biology: if something no longer sparks functionality, it's out. But autophagy declines with age. The formation of autophagosomes—the structures that ferry cellular junk to the lysosome for destruction—becomes sporadic and inefficient. The result? Cellular waste accumulates.

Start with the brain.

Neurons, unlike many other cells, don't regularly divide. They have to last a lifetime. So they're heavily reliant on pristine proteostasis. But with age, this balance tilts. Misfolded tau proteins and beta-amyloid peptides begin to aggregate into insoluble plaques and tangles. These clumps aren't just biochemical noise—they physically disrupt cell communication, choke synapses, trigger inflammation. In Alzheimer's disease, this accumulation becomes pathology. But even in healthy aging, the brain shows signs of protein debris.

Move to the muscles.

Aging muscles lose mass and strength—a condition called sarcopenia. Some of that is due to hormonal changes and mitochondrial decline. But a crucial contributor is proteostasis failure. Studies show that aged muscle fibers accumulate oxidized, cross-linked, or misfolded proteins. The cleanup machinery slows, and protein turnover—the process by which old proteins are replaced with new—is delayed. The result? Stiff, less responsive tissue. The machinery meant to keep us mobile gums up with molecular soot.

Now look at the lens of the eye.

Cataracts form when crystallin proteins in the lens—normally transparent and perfectly ordered—begin to clump and lose their solubility. These proteins are meant to last decades without replacement. But without effective proteostasis, they degrade. Light scatters. Vision clouds.

Cataracts aren't just a defect of the eye—they're a window into cellular housekeeping gone awry.

This pattern holds across systems. Wherever proteins must be folded, transported, used, and degraded with precision—meaning, everywhere—the loss of proteostasis spells dysfunction.

But how does this happen?

Part of the answer lies in the mounting burden on the system. As we age, we're exposed to more stress: oxidative damage from mitochondria, glycation from excess sugar, UV radiation, chronic inflammation. Each of these factors increases the load of damaged proteins. Meanwhile, the cleanup crew shrinks. Chaperone production declines. Proteasomes lose efficiency. Autophagy slows. The cell is asked to clean more, faster, with fewer tools.

Another factor is transcriptional noise. The genes that encode for cleanup machinery are themselves regulated epigenetically—by chemical tags and chromatin accessibility. As we saw in the previous chapter, epigenetic drift with age can downregulate essential genes, including those for proteostasis. The orchestra starts misreading the music.

And there's the feedback problem. Proteins that aren't cleared can bind to and damage other proteins. This misfolding can spread in a prion-like manner—corrupting what was once functional. It's not a static breakdown. It's a spreading fire.

Still, there's a tension in the field. Not all aggregates are inhcrcntly bad.

In recent years, some researchers have questioned whether the cell's tendency to sequester misfolded proteins into aggregates might be protective. Better to isolate toxic elements into stable clumps than let them roam freely. In this view, plaques and tangles aren't malfunctions; they're triage decisions. It's a view that complicates the picture, but it makes sense. Biology often chooses the least worst option.

This duality—damage vs defense—is echoed in hormesis, the idea that low levels of stress can be beneficial. Mild heat shock, caloric restriction, or intermittent fasting can boost proteostasis by temporarily stressing cells into cleaning up. The garbage strike ends only when the city gets fed up and sends reinforcements.

AGGREGATES AND CATASTROPHES

Out in the Pacific Ocean, far from the reach of any nation, there's a place where the water seems calm but is anything but clean. It's called the Great Pacific Garbage Patch—a vast soup of microscopic plastic particles and larger debris, suspended not at the surface, but swirling beneath, invisible to satellites, often ignored by shipping routes. For decades, no one noticed. A plastic bottle tossed in Tokyo, a fishing net cut loose off the coast of California—these scattered decisions created no ripple. But eventually, the garbage found each other. The currents carried them into gyres, great circular loops of ocean flow, where time had one goal: accumulation. What began as fragments grew into islands. And what once could've been cleaned up by hand now over-whelms ecosystems. Fish mistake it for food. Coral suffocates beneath it. Entire swaths of ocean biology are altered by the weight of what didn't belong.

Inside the human body, something eerily similar unfolds.

Unlike broken DNA, which sits quietly and triggers repair mechanisms, protein aggregates are chemically and physically invasive. They don't just sit—they spread. They seed. They behave, in many ways, like biological weeds. One seed crystal of misfolded amyloidbeta can nucleate a chain reaction that builds the infamous plaques in Alzheimer's disease. A single misfolded alpha-synuclein can initiate the clumping that leads to Lewy bodies in Parkinson's. These aggregates aren't just byproducts—they're active disruptors, gumming up the cellular gears, clogging the synaptic networks, inducing oxidative stress, and triggering inflammation.

Most readers will be familiar with Alzheimer's or Parkinson's. But the aggregation story goes far beyond the brain. Cataracts, for instance, result from clumping of crystallin proteins in the eye's lens. The proteins there are unique—they're formed early in life and never replaced. Over time, exposed to sunlight, oxidative stress, and environmental wear, they denature and stick together, scattering light instead of focusing it. The world blurs not because the eye is tired, but because its transparency has been invaded by a slow, silent flood of accumulated error.

For decades, researchers believed that these aggregates were simple markers—passive indicators of decline. But recent work has shown they can act like prions, proteins that misfold and then induce others to do the same, like a contagious bad example. One misbehaving molecule can turn a dozen others. It's a lesson in biological sociology: even proteins succumb to peer pressure under the wrong conditions.

And perhaps most chillingly, we now know that these aggregations don't always stay local. In Alzheimer's disease, aggregates form in specific brain regions—but over time, they spread, moving along neural circuits like a slowburning fuse. From the hippocampus to the cortex, the misfolded proteins march, eroding memory, cognition, and identity in their wake.

Not unlike the expanding gyre in the Pacific, their trajectory is slow but relentless.

What's striking—and what makes the aging process so complex—is that these catastrophic accumulations don't require a new invader. They don't depend on viruses or environmental toxins. They arise from within, from the ordinary molecular turnover of life. It's as if the very act of living plants the seeds of decline. And while young cells clean up the messes efficiently, older cells become bogged down, increasingly unable to distinguish between salvageable parts and junk.

The metaphor of the Great Pacific Garbage Patch is apt not just for its scale, but for its invisibility. Aggregates start small. They begin with a folding error. A single protein missed by the chaperones. A minor slowdown in the autophagy machinery. No crisis. No symptom. But slowly, as the cellular current swirls, the trash accumulates. It resists degradation. It fragments into microdebris. It spreads.

And one day, the system—whether marine or molecular—no longer functions as it once did.

There is a tragic irony here. The proteins most vulnerable to aggregation are often the ones most essential: those involved in memory, movement, clarity, vision. Aging, then, isn't merely about time—it's about accumulation. And like the ocean, our cells have no infinite capacity for waste.

Yet, unlike the Pacific, the body retains some measure of renewal. In some cases, researchers have found ways to stimulate the cellular garbage collectors—to reawaken autophagy or enhance proteasome activity. Drugs, dietary stress, and genetic tools are being explored to reduce the garbage load. The cleanup is difficult, but not impossible. Still, in most people, the damage builds faster than it can be cleared.

This part of the story doesn't end in dramatic collapse. There is no tsunami of protein chaos. There is, instead, a slow silting—a quiet erosion of function. The hand that shakes. The voice that forgets. The skin that sags not from vanity, but from biomechanical truth.

WHEN STRESS CLEANS YOU UP

In the towering forests of the western United States, destruction is often the start of rebirth. Every few decades, a forest fire will tear through thousands of acres—scorching trees, blackening soil, sending plumes of smoke into the sky so thick it turns day into dusk. From a distance, it looks apocalyptic. The charred skeletons of trees, the silence where birdsong used to echo, the heatwarped landscape—all of it feels like loss.

But that's not how nature sees it. Fire, for all its ruin, is also renewal. Beneath the ash, seedlings begin to stir. Grasses push through the blackened crust. And some pinecones—sealed shut for decades—only open when kissed by flame. These are serotinous cones, nature's fire vaults, which release their seeds only in response to high temperatures. Fire doesn't just destroy. It clears the clutter, returns nutrients to the soil, and resets the balance. Without it, forests become choked with deadwood, fragile under-brush, and tangled decay. What seems like devastation is actually a kind of necessary purge—a built-in resilience mechanism evolved over millennia.

Our cells operate on a similar principle.

A little stress, strategically applied, doesn't break the body—it sharpens it. This is the core of hormesis: the idea that mild, temporary challenges can trigger adaptive responses that leave the organism stronger than before. Just as fire teaches the forest to grow back better, cellular stress teaches the body to clean itself up.

At the heart of this cleanup crew is autophagy—literally "self-eating." It sounds cannibalistic, but it's one of the most elegant forms of cellular recycling ever devised. When cells detect low energy or damage, they don't panic. They pivot. They wrap wornout organelles and misfolded proteins in membranes and funnel them into lysosomes—the acidic digesters of the cell. Inside, the biological junk is dismantled into parts that can be reused. Nothing goes to waste. This isn't desperation—it's efficiency. During fasting, autophagy ramps up. During exercise, it kicks into gear. Even heat or cold exposure can do it. Cells interpret these stressors not as threats, but as prompts to tidy up.

Consider what happens when autophagy is suppressed. In older animals—mice, worms, and humans alike—this internal recycling slows. Waste builds up. Damaged proteins linger in the cytoplasm like uncollected trash. Energy production falters. Immune cells become sluggish. But if we can reignite this system, even briefly, the results can be astonishing.

In one study, mice genetically engineered to boost autophagy lived longer and stayed healthier, maintaining better muscle function and metabolic control late into life. In another, periodic fasting activated autophagy so effectively that older mice showed improved organ function and resistance to age-related disease. Even in humans, short-term fasting has been linked to enhanced immune responses, in part because of its ability to clear out damaged white blood cells and replace them with fresh ones. It's as if the body, when denied constant input, turns inward to restore itself.

But how did we even discover this miraculous cleaning system?

The answer starts with an accident.

In the 1960s, a group of scientists in a genetics lab in Italy were studying fruit flies. It was the peak of summer, and an unrelenting heat wave had pushed the lab's temperatures far beyond the norm. Flies that were thriving one day suddenly began producing strange proteins the next. At first, the researchers thought it was a malfunction in the experiment. But upon closer inspection, they realized the flies were responding to heat stress by manufacturing a set of entirely new proteins—ones that helped stabilize their internal systems and protect them from damage. These were the first "heatshock proteins," later found in virtually every organism from bacteria to humans.

Heatshock proteins are chaperones. Not in the high school dance sense, but in the molecular one: they help other proteins fold correctly. And when cells are under stress—be it thermal, oxidative, or nutritional—these chaperones rush to the scene like emergency responders, stabilizing the wobbly structures of vulnerable proteins, giving them a second chance at proper function. If folding fails, they help tag the protein for degradation. They don't panic either. They triage. They manage chaos.

This ancient machinery is still hardwired into us. Every time we sweat through a workout, fast through a meal, or take a brisk cold shower, we're reminding our cells that it's time to adapt. We're creating a controlled burn—not enough to harm, but enough to signal that it's time to clean up the deadwood. This isn't masochism. It's biology.

Hormesis reframes how we think about aging. It's not just about avoiding damage; it's about managing how we respond to it. Most of our modern environment is designed to minimize stress—climate control, convenience food, endless entertainment, smooth roads, and ergonomic chairs. We've insulated ourselves from discomfort. But in doing so, we may have deprived our cells of the very cues they need to stay sharp.

Even the most cuttingedge therapies in longevity research—like compounds that mimic fasting or activate autophagy—are, at their core, imitating what nature already figured out. Caloric restriction, once seen as fringe, is now a pillar of aging research. Compounds like spermidine, found in fermented foods, or rapamycin, derived from soil bacteria, are being studied for their ability to trigger cleanup pathways. All of these essentially light small, controlled fires inside the cell. They don't destroy. They rejuvenate.

Back in the biotech startup, the young researcher meets with the chef 's family. She doesn't offer miracle cures. But she speaks with cautious optimism. "There's growing evidence," she says, "that certain diets, specific types of movement, even mild thermal stress—like sauna exposure—can enhance the body's ability to clear out misfolded proteins. We can't stop all the damage, but we might be able to clean it up better."

The family nods. There's no magic bullet. But there is something beautiful in the idea that renewal doesn't have to come from outside. It can be triggered from within. They begin small—morning walks, lighter dinners, longer breaths in the sunshine. The changes are modest, but the intention is profound: to create little sparks of stress that might rekindle a deeper cellular flame.

The forest, after fire, is never the same. But in many ways, it's stronger. The trees that rise from the ash are more resilient, their roots deeper, their canopies more open to the sky. So too, perhaps, can we be—not by avoiding life's fires, but by letting them teach us how to grow back better.

WHAT TO KEEP, WHAT TO CLEAR

In the heart of Nara, Japan, stand wooden temples that have outlived empires. The wood beneath your feet may be over a thousand years old—weathered by earthquakes, soothed by monks' sandals, and polished by time itself. These aren't lifeless relics. They breathe. And they're still cared for.

Master carpenters—known as miyadaiku—have maintained these temples for generations. But their approach is strikingly different from what you'd find in modern repair. They don't demolish and rebuild. They listen. They knock gently on each beam, learning which parts have rotted and which are still sound. They sand away only what must go. They oil and reinforce what endures. And if something breaks, they may practice kintsugi—repairing it with gold lacquer that draws attention to the fracture rather than hiding it. The wound becomes part of the beauty. The philosophy is simple: not all damage should be erased. Some should be honored.

Inside the aging human body, a similar choice must be made every day: what to keep, and what to clear.

We've spent the chapter exploring proteostasis—the intricate system our cells use to fold, monitor, and dispose of proteins. But aging disrupts this balance. As cells grow older, their chaperones lose vigilance, their trash cans over-flow, and their cleanup crews falter. The result isn't just a pileup—it's confusion. Some proteins are misfolded but harmless. Others are toxic but hard to recognize. Some clumps may even be protective, a desperate act of self-containment like taping off a gas leak. The line between what should stay and what must go becomes harder to draw.

This is the core dilemma of latestage proteostasis failure: aging isn't just an accumulation of trash—it's a failure to discriminate. And that makes cleanup much harder.

On one side of the balance is the overwhelming load. Misfolded proteins, oxidized debris, cross-linked molecules—all accumulating faster than the cell can manage. On the other side is the dwindling capacity to clear them: sluggish autophagy, distracted chaperones, and overburdened proteasomes. Aging is what happens when that balance tilts too far, too long. But even in the face of this slow tipping, science is beginning to nudge the scales back.

One of the most promising nudges comes from spermidine, a compound found in aged cheese, mushrooms, soy products, and even wheat germ. It's been shown in animal studies to activate autophagy, improve cognitive function, and extend lifespan. The name might be unfortunate, but the effects are worth noting: spermidine doesn't just "clean"—it trains the body to resume its own housework.

Then there's rapamycin, discovered on Easter Island, originally used as an immunosuppressant. When administered at low doses in mice, it extends lifespan by activating the mTOR pathway—essentially tricking cells into a fasting-like state where they ramp up internal maintenance. Human trials are now underway, cautiously. But early signs point to rapamycin as a drug that might rejuvenate not just cells, but their systems of self-repair.

Still, the most compelling interventions may not come in pill form at all. Intermittent fasting, already popularized for weight loss, has been shown to induce autophagy and improve markers of aging in both animals and humans. Heat therapy—like sauna use—activates heat shock proteins and promotes cellular stress resilience. Cold exposure appears to do the same, albeit via different molecular triggers. These aren't newage fads; they're ancient biological cues our bodies still respond to, if we let them.

Yet with every advancement comes a note of caution. In Alzheimer's research, decades were spent targeting amyloid plaques—the sticky protein aggregates that litter the aging brain. Dozens of drugs were developed to remove them. Many succeeded. But the patients didn't always improve. Removing the plaque didn't reverse the dementia. And that led scientists to a disturbing hypothesis: maybe those plaques weren't the enemy. Maybe they were a desperate containment effort. A cellular act of triage.

In other words, the body may sometimes choose sequestration over removal. When clearance mechanisms fail, and toxicity threatens, it might be better to quarantine a threat than to battle it endlessly. This is the nuance of aging: it's not always about dysfunction. It's about tradeoffs.

The same tradeoffs are seen in nature. An old temple beam may carry cracks. But if its core is strong, a master doesn't replace it—he reinforces it. The crack becomes part of the design. The flaw is embraced. This isn't sentimentality. It's wisdom. Total restoration isn't always necessary. Sometimes, partial integrity is enough to preserve function. In the same way, an aging body doesn't need to become perfect—it needs to become functional enough to keep going.

The key is knowing what to keep and what to clear. Not all damaged proteins are villains. Not all aggregates are death sentences. And not all cleanup is curative. Aging is a negotiation between damage and adaptation, loss and reinvention.

That brings us back to our chef.

He's no longer in charge of the kitchen. The pace was too demanding, the tremors too unpredictable. But retirement, for him, wasn't an end. It was a refocus. These days, he teaches a small group of students in a community kitchen. The sessions are quiet, deliberate. He speaks slowly. He tastes more than he chops. And yet, in these unhurried lessons, something has returned.

One afternoon, mid-lesson, he picks up a whisk. His hands, for a moment, move with forgotten fluidity. Muscle memory wakes up. He smiles—not at the dish, but at the motion. A flicker of former grace, unburdened by outcome. It passes quickly. But it lingers, too.

The kitchen is not spotless. Some mess remains. But it's no longer chaos. The tools are within reach. The counters are familiar. The heat of the stove is steady. It's not the restaurant of his youth—but it is still a place of warmth and nourishment. The house, still, is in order.

And that, perhaps, is the essence of healthy aging—not perfection, but maintenance. Not endless youth, but careful, loving repair. To age well is to know what to let go, and what to fold gently back into the shape of life.

Reader-tool boxes for this chapter

CLOSER LOOK ***How a cell folds a protein — and what helps when it fails***

A protein begins life as a linear string of amino acids extruded from a ribosome. Within milliseconds of emerging, it must fold into a precise three-dimensional shape determined by its sequence. For small proteins, the folding is spontaneous. For larger or more complex ones, the cell deploys molecular chaperones — Hsp70, Hsp90, the chaperonin TRiC/CCT — that bind hydrophobic patches and prevent aggregation while folding completes.

When folding fails, the cell has options. It can refold the protein with help from chaperones. It can tag it with ubiquitin and feed it into the proteasome for degradation. Or it can engulf it in an autophagosome and ship it to the lysosome to be dissolved. Aging compromises all three pathways. Chaperone capacity declines, proteasome activity drops, and autophagic flux slows — leaving misfolded proteins to accumulate and aggregate.

Hartl, Bracher, & Hayer-Hartl (2011). Molecular chaperones in protein folding and proteostasis. Nature, 475(7356), 324–332.

EASY ACCESS ***Autophagy in one minute***

Imagine your cell as a small city. It runs on machinery. Machinery breaks. Trash piles up. Autophagy is the city's recycling truck — a process where the cell wraps damaged proteins and worn-out organelles in a membrane sac, ships them to the lysosome (a vat of digestive enzymes), and breaks them back down into raw materials. Those raw materials get reused. With age, the truck makes fewer rounds. Trash accumulates. Function suffers. Most of the longevity interventions you'll meet in the next chapter — fasting, rapamycin, exercise — work in part by sending more recycling trucks out on the road.

CLOSER LOOK *When proteins aggregate — the geometry of disease*

Most age-related neurodegenerative diseases share a common architecture. A normally soluble protein adopts a misfolded shape rich in beta-sheet structure. That shape is sticky to itself. One misfolded protein recruits another, which recruits another, building first into oligomers, then into larger fibrils, and ultimately into the macroscopic plaques and tangles visible in autopsy. Beta-amyloid in Alzheimer's, alpha-synuclein in Parkinson's, TDP-43 in ALS, mutant huntingtin in Huntington's — different proteins, same physics.

The amyloid hypothesis of Alzheimer's, now in its third decade, holds that imbalance between production and clearance of beta-amyloid 42 is the upstream initiating event. The hypothesis has weathered controversy and partial vindication — recent anti-amyloid antibodies have, for the first time, modestly slowed cognitive decline in trials, suggesting that clearance matters even late in the disease.

Selkoe & Hardy (2016). The amyloid hypothesis of Alzheimer's disease at 25 years. EMBO Molecular Medicine, 8(6), 595–608.

CLOSER LOOK ***Why a missed meal helps the cleanup crew***

Autophagy is regulated by the cell's two great nutrient sensors. mTOR (mechanistic target of rapamycin) reads abundance — when amino acids and growth signals are plentiful, mTOR fires, telling the cell to grow and suppressing autophagy. AMPK reads scarcity — when ATP runs low, AMPK fires, telling the cell to conserve and inducing autophagy. Caloric restriction, fasting, and exercise all push this balance toward AMPK and away from mTOR.

Beyond timing, certain compounds nudge the same dial. Rapamycin directly inhibits mTOR. Spermidine, a polyamine present in foods like wheat germ and aged cheese, induces autophagy in yeast, worms, flies, and mice — and observational data in humans suggest higher dietary intake correlates with lower all-cause mortality. None of this is a prescription; it is the signature of a system that responds to what the cell has been told about its environment.

Eisenberg et al. (2009). Induction of autophagy by spermidine promotes longevity. Nature Cell Biology, 11(11), 1305–1314.

KEY TERMS ***Vocabulary of the proteostasis chapter***

Proteostasis — The cell's overall protein quality control system: synthesis, folding, trafficking, and degradation.

Chaperone — A protein that helps other proteins fold correctly. Hsp70 and Hsp90 are the major aging-relevant ones.

Proteasome — A barrel-shaped enzyme complex that chops up tagged, damaged proteins.

Autophagy — Self-eating. The lysosomal pathway by which the cell recycles damaged components.

Aggregate — A clump of misfolded protein. The structural feature shared across most age-related neurodegenerative diseases.

Lysosome — Acidic, enzyme-filled organelle that digests cellular debris into reusable parts.

THE LONGEVITY DIET

DEREGULATED NUTRIENT SENSING

It wasn't what she expected. There were no candles, no savory aroma of roasted vegetables or rich broths wafting through the air. The room, a community center in Palo Alto, smelled faintly of green tea and anti-septic wipes. The journalist had imagined a kind of healthconscious potluck—maybe chia puddings, quinoa salads, lentil stews. Instead, what sat before her was a group of middle-aged people dissecting a handful of steamed vegetables, half an avocado, and a single Brazil nut with the reverence of monks performing a sacred rite.

These were members of the Calorie Restriction Society, and this was their monthly shared meal. They were lean, almost preternaturally so, with sharp cheekbones and the kind of alert, unblinking eye contact that suggested their bodies had adapted to running on fumes. One man wore a blood glucose monitor. A woman checked her ketone strips before eating. There was no bread. No seconds. No dessert. When the journalist asked how they could enjoy this, someone said, "This isn't about enjoying dinner. This is about enjoying 100 future birthdays."

They weren't joking.

This peculiar ritual is the front line of one of the most studied, controversial, and surprisingly robust anti-aging interventions known to science: caloric restriction. Long before the mechanisms were understood, researchers noticed that lab animals fed less lived longer. In 1935, a Cornell scientist named Clive McCay reduced calorie intake in rats by about 30% and observed not just a dramatic extension in lifespan but a delay in the signs of aging. These rodents had shinier coats, fewer tumors, and moved with youthful energy well into their senior months.

For decades, the effect remained mysterious. Was it simply the reduction in oxidative damage from fewer metabolic reactions? Or was something deeper at play—some systemic switch that told the body to stop growing and start repairing?

The problem is: we don't face famine anymore. Not in the affluent world. Most of us live in a state of nutritional abundance bordering on overload. Constant availability of food, often high in sugar and protein. We're stuck in grow mode. Repair is postponed indefinitely. Cells keep dividing, but they accumulate damage. Cancer risk rises. The cellular garbage piles up. Eventually, growth becomes pathological.

This, in essence, is the idea behind deregulated nutrient sensing as a hallmark of aging. In youth, these nutrientsensing pathways are vital. They help us grow, reproduce, and respond to injury. But with age, their relentless activity becomes toxic. They don't shut off when they should. The balance between growth and maintenance tips too far in one direction. What helped us survive the first part of life hastens our decline in the second.

In the room with the calorie restrictors, the journalist watched as a man in his fifties slowly peeled the skin off a tomato slice, explaining that even the small glycemic bump mattered to him. He had the fasting insulin levels of a teenager. His cholesterol, blood pressure, and inflammatory markers were textbook perfect. He hadn't had a cold in years. And yet, he looked... brittle. Not ill, but not entirely well. She wondered: could there be a cost to this nearascetic way of life?

TOR AND THE COST OF CONSTANT GROWTH

In the hills of central Japan, tucked behind pine forests that shed mist in the early morning, there exists a monastery known not for its spiritual guidance but for something more peculiar—its silence. This temple has no chants, no bells. Its monks speak only through gestures. But what makes it truly exceptional is their garden. It contains a tree, planted five hundred years ago, that still bears fruit.

The monks believe the secret lies in restraint. They prune the tree with almost obsessive delicacy, never over-watering, never over-feeding. Growth is not the goal—survival is. One monk once explained to a visiting botanist, using only gestures and a small blackboard, that the tree lived long because it was never forced to grow too fast. "It listens to the seasons," he wrote.

That ancient orchard serves as an elegant metaphor for a molecule most people have never heard of: mTOR. Mechanistic Target of Rapamycin. It's the conductor of the cellular orchestra, deciding whether cells should divide, grow, or hold off. And in the story of aging, mTOR is the central figure that keeps whispering the same command to our cells: grow now, fix later.

But cells that always grow never get around to fixing. In its youth, the human body is an explosion of anabolic ambition. mTOR is heavily active—driving muscle building, reproduction, and cell division. It senses amino acids and energy levels, and when it feels abundance, it signals the machinery to build. That's great when you're building a body. But what happens when that body is already built?

Aging, as it turns out, may be partly due to the fact that mTOR doesn't know when to shut up. Or more precisely, the environment never tells it to. In a world of constant feeding—three meals a day, snacks in between, protein powders and midnight fridge raids—mTOR stays on. The signals that would usually downregulate it, like fasting or nutrient scarcity, rarely arrive.

And that's where the trouble begins.

Because mTOR doesn't just promote growth. It inhibits autophagy—the self-cleaning process that cells rely on to clear out damaged proteins, organelles, and molecular debris. Think of autophagy as a biological decluttering system. It's like what the monks do to their orchard: pruning dead branches to save the living ones.

In a constantly fed state, autophagy is suppressed, and cellular clutter accumulates. Damaged mitochondria, oxidized proteins, and misfolded molecules are all left behind, slowly sabotaging the cell from within. The result is not explosive illness, but slow degradation. Tissues lose resilience. Organs falter. The biological machinery, once perfectly tuned, begins to jam.

The story of mTOR took a dramatic turn in 2009, when researchers at the Jackson Laboratory did something no one expected. They gave old mice—20-month-old mice, equivalent to humans in their 60s—a drug called rapamycin. Derived from soil on Easter Island, rapamycin inhibits mTOR. It was originally developed as an antifungal compound, then as an immunosuppressant for organ transplants.

But when these elderly mice were fed rapamycin, they lived longer. Much longer.

This was the first time in history that a pharmacological agent extended the lifespan of mammals even when introduced late in life. The implications were staggering. It meant that even aging cells, already groaning under the weight of years, could be tricked into acting young—if only the growth signals could be muted.

In other words, the secret wasn't in adding more—it was in turning something down.

The longevity world took notice. Suddenly, mTOR wasn't just a footnote in molecular biology—it was a celebrity. Researchers began to see aging not simply as a process of damage, but of mistaken priorities. Cells were acting as if they were still in growth mode, when what they needed was repair. Like a city building new skyscrapers while ignoring the cracked foundations beneath them.

The insight rippled into everyday life. Athletes obsessed with high-protein diets began questioning whether muscle gain came at the cost of long-term health. Tech billionaires started intermittent fasting routines and funded companies testing rapamycin in humans. Meanwhile, deep in academic journals, the nuance began to emerge.

mTOR isn't bad. It's context-dependent.

The monks' tree isn't strong because it lacks growth. It's strong because its growth is timed. mTOR, too, must dance to a rhythm—on when needed, off when not. The problem in modern life is that we've drowned out the off switch. Our constant availability of nutrients has broken the seasonal cycle that evolution embedded into our physiology. We live in a perpetual harvest.

But longevity, it seems, favors winter.

In aged tissues, mTOR stays active. It drives cells to keep growing even when they are damaged. In some cases, that promotes cancer. In others, it causes hypertrophy without repair—bigger, weaker cells that accumulate damage faster. In muscle, this contributes to sarcopenia; in the brain, it can lead to loss of resilience. That's why suppressing mTOR can improve not only lifespan but healthspan.

Yet here lies the paradox. Inhibition of mTOR may promote longevity, but it can also impair immunity and wound healing. Like fire, it must be managed, not extinguished.

The key, then, is timing. Fasting windows. Nutrient pulses. Lifestyle patterns that gently toggle mTOR between action and rest. Not constant suppression, but modulation. Just as the monks know when to prune and when to let the tree flower.

WORMS THAT WOULDN'T DIE

In the basement of a lab at the University of California, San Francisco, Cynthia Kenyon stared at a petri dish of tiny worms. They were C. elegans, about the size of a comma on a page. Most lived just a few weeks—simple lives, simple deaths. She had seen it dozens of times. But something was different this time. These worms weren't dying.

They weren't just surviving; they were thriving—squirming with the same energy as the day they were born. Days passed. Then weeks. The control group withered, as expected. But this batch? They just kept going.

The gene was daf-2, part of the insulin/IGF-1 pathway. And that one tweak—like deleting a sentence from a book—doubled the worm's lifespan.

The implications landed like an earthquake.

Until that moment, aging was assumed to be inevitable. Like gravity, or taxes. But Kenyon's worms weren't breaking—they were simply refusing to decline. The cells weren't in denial about aging. They had just stopped getting the signal to rush toward it.

That petri dish was where the modern science of nutrient sensing and aging began.

The insulin/IGF-1 pathway isn't glamorous. It doesn't have a poetic name like "fountain of youth." It's a basic, ancient pathway that evolved in almost every animal to detect one thing: is there food?

When nutrients flood in—especially sugars and proteins—this pathway flips on. The body takes the signal to mean: time to grow, time to reproduce, time to thrive. When the signal dims, the body switches gears: time to repair, to recycle, to conserve.

Kenyon's mutation essentially turned down this pathway. The worms still functioned, but their biology leaned toward maintenance instead of growth. Remarkably, they didn't just live longer—they lived better. They aged more slowly, delayed disease onset, and showed higher resilience to stress. As Kenyon put it, "It wasn't about stretching out decline—it was about delaying the start of it."

She began testing more genes in the same pathway. The results echoed. More tweaks, more long-lived worms. The pattern repeated in other organisms—fruit flies, yeast, even mice. Knock down insulin/IGF signaling, and aging slowed. It didn't matter the species.

A unifying theme emerged: overactive nutrientsensing made organisms burn bright but fade fast. Dial it down, and the system endured.

Then came the human connection—unexpected and deeply revealing.

In Ecuador, scientists had been following a small population of individuals with Laron syndrome, a genetic disorder that results in unusually low levels of IGF-1. These individuals are short in stature—none taller than four feet—but strikingly resistant to two things: cancer and diabetes. Despite obesity and poor diets, they rarely developed either condition.

What's more, these individuals exhibited markers of slowed cellular aging: lower oxidative stress, reduced DNA damage, and better insulin sensitivity.

It was as if nature had already run the experiment.

Some researchers now suspect that populations with naturally lower IGF-1 activity might harbor hidden advantages—longevity, resistance to metabolic disease, even cognitive protection in old age.

The science went deeper. IGF-1 (Insulin-like Growth Factor 1) doesn't act alone. It works in concert with insulin itself, and both are activated in response to food—particularly sugar and protein. High IGF-1 drives growth. That's great for children. But in adults, excessive IGF-1 keeps cells dividing when they should be repairing.

That constant cellular division increases the chance of errors—mutations, misfolded proteins, metabolic strain. It's the biological equivalent of running your car engine at full throttle, all the time.

Lowering IGF-1 signaling doesn't shut the system down. It simply tells the cells: slow down, check your work, clean up as you go. And in that slower pace, health is preserved.

This is why calorie restriction has such profound effects in animals. It lowers IGF-1. But the same effect can be achieved through timing—fasting, protein cycling, even reducing certain amino acids. It's not just how much you eat; it's how often and what kind.

In one experiment, researchers fed mice a low-protein diet that specifically reduced methionine, an amino acid that activates IGF-1 and mTOR. The mice lived longer, with fewer tumors. Other studies showed that intermittent fasting suppressed IGF-1 without reducing total calorie intake. The takeaway: we can modulate nutrientsensing not just with quantity, but quality and rhythm.

And yet, there's nuance. Some IGF-1 is necessary. It supports muscle mass, bone health, and brain function. Athletes, for instance, may benefit from slightly higher IGF-1—while balancing it with periods of recovery and fasting.

Cynthia Kenyon's worms rewrote the assumptions. Her work sparked a global race to understand nutrient sensing, culminating in drugs like metformin and rapamycin, diet strategies like fasting and protein cycling, and a deeper appreciation that aging is plastic—changeable, manipulatable.

CLOSER LOOK *The two amino acids that tell mTOR the table is set*

Not every amino acid speaks to mTOR with the same voice. Two — leucine and methionine — do most of the talking, and the cell listens carefully to both.

Leucine is sensed by a protein called Sestrin2, which sits like a doorkeeper on a larger complex called GATOR2. When leucine is scarce, Sestrin2 holds GATOR2 closed and mTORC1 stays quiet. When leucine binds — and it binds with striking precision, at concentrations easily reached after a protein-rich meal — Sestrin2 releases GATOR2, and mTORC1 fires. The cell takes this as a signal that building materials have arrived and shifts into anabolic mode: more protein synthesis, more growth, less autophagy.

Methionine binds the same sensor, less tightly but by the same mechanism, adding its own vote to the chorus. Together these two amino acids function as the cell's protein-meal detectors. The implication is unsettling for anyone who treats protein as uniformly virtuous.

Leucine is concentrated in red meat, whey, and dairy; methionine is concentrated in eggs, meat, and fish. Diets dominated by these foods keep mTORC1 turned up, day after day, and the cleanup crew that runs only when mTORC1 quiets down rarely gets its shift. In mice, simply restricting dietary methionine — without cutting calories — extends maximum lifespan and lowers IGF-1, insulin, glucose, and thyroid hormone in patterns that mirror caloric restriction. The lesson is not that protein is dangerous. It is that the cell reads the amino acid composition of what you eat, and certain amino acids speak louder than others.

Levine, M. E., Suarez, J. A., Brandhorst, S., Balasubramanian, P., Cheng, C.-W., Madia, F., Fontana, L., Mirisola, M. G., Guevara-Aguirre, J., Wan, J., Passarino, G., Kennedy, B. K., Wei, M., Cohen, P., Crimmins, E. M., & Longo, V. D. (2014). Low protein intake is associated with a major reduction in IGF-1, cancer, and overall mortality in the 65 and younger but not older population. Cell Metabolism, 19(3), 407–417.

Miller, R. A., Buehner, G., Chang, Y., Harper, J. M., Sigler, R., & Smith-Wheelock, M. (2005). Methionine-deficient diet extends mouse lifespan, slows immune and lens aging, alters glucose, T4, IGF-I and insulin levels, and increases hepatocyte MIF levels and stress resistance. Aging Cell, 4(3), 119–125.

Wolfson, R. L., Chantranupong, L., Saxton, R. A., Shen, K., Scaria, S. M., Cantor, J. R., & Sabatini, D. M. (2016). Sestrin2 is a leucine sensor for the mTORC1 pathway. Science, 351(6268), 43–48.

EASY ACCESS ***What your plate tells your cells***

Every meal sends a message. When that message is heavy in animal protein — a steak, a stack of scrambled eggs, a whey shake — it's especially rich in two amino acids called leucine and methionine.

Inside your cells, a sensor reads those amino acids almost the way a thermostat reads heat.

Plenty of leucine?

The cell flips into build-mode: more protein synthesis, more growth, less self-cleaning.

Scarce leucine and methionine?

The cell flips into maintenance: less growth, more recycling of old and damaged parts. Neither setting is "good" or "bad" — both are needed at different times of life.

The problem comes when the build-mode message is sent every single meal, every single day, for decades. The cleanup crew rarely clocks in. Cells accumulate worn-out parts. Plant proteins generally send a softer version of the same message, which is one reason researchers find that the same gram of protein from beans and lentils behaves differently in long-term health data than the same gram from beef. The takeaway isn't to fear protein. It's to notice that what kind of protein, and how often, are the variables your cells are actually reading.

SIGNALS IN THE SILENCE

In the pre-dawn stillness of the Kalahari Desert, a hunter rises. There's no breakfast, no coffee, no hurried spoonful of oats. Just the wind scraping across dry sand and the faint imprint of hoofprints heading east. Hunger is not a distraction—it is a state of heightened awareness. He moves lightly, silently, tracking an antelope across the dunes. His muscles are lean, his breath steady, his mind clear. No food, but no fatigue. Evolution didn't make fasting a weakness. It made it a trigger.

Across millennia, humans have endured cycles of feast and famine, bounty and barrenness. Our ancestors didn't die from missing a meal; they survived because their bodies were hardwired to adapt. In those stretches of scarcity, something remarkable would happen inside their cells—a shift from consumption to conservation, from growth to resilience.

At the molecular level, this shift was orchestrated by pathways most of us have never heard of: AMPK and sirtuins. These aren't just backup systems—they're survival programs, activated not by excess but by absence.

AMPK—short for AMP-activated protein kinase—is a kind of internal fuel gauge. It constantly measures the ratio of energy molecules inside a cell. When you eat, that ratio stays high. When you don't, the balance tips: ATP (cellular energy) drops, AMP rises. That's the moment AMPK steps in.

Activated by energy scarcity, AMPK tells the cell to stop burning resources and start conserving. It inhibits processes that require energy—like fat production—and flips on pathways that generate it—like fat burning and glucose uptake. It's the metabolic equivalent of turning off the lights and heating only one room in a house.

But AMPK does more than pinch pennies. It initiates autophagy, the cell's self-cleaning process. Junk proteins, damaged mitochondria, malfunctioning components—AMPK signals the cell to recycle them. It's a cleanup crew activated by the simple state of not eating.

This is why fasting, even short-term, has effects that far exceed weight loss. It doesn't just change your weight—it changes your metabolism, repair processes, and cellular priorities.

Running alongside AMPK is a more enigmatic partner: sirtuins. Named after the yeast gene sir2 (silent information regulator), sirtuins are a family of proteins that respond to energy stress and modulate gene expression. They're the switches that turn on longevity programs when the body senses a lack of nutrients.

Sirtuins are NAD^+-dependent, meaning they require nicotinamide adenine dinucleotide, a molecule that rises during fasting and exercise. When activated, sirtuins reinforce DNA repair, bolster mitochondrial function, and increase stress resistance. They don't just help cells survive—they help them age better.

This led to what became one of the most captivating—and controversial—stories in modern longevity science: resveratrol, the compound in red wine that appeared to activate sirtuins.

The headlines came fast: \"Red wine extends life!\" \"Drink your way to youth!\" The idea was intoxicating—literally. If a glass of Pinot Noir could mimic fasting and exercise, perhaps aging could be delayed without sacrifice.

But science is rarely so convenient. The reality? Resveratrol did show promise in lab models—yeast, worms, even some mice. But the effects were dose-dependent and often inconsistent. To match the levels used in some animal studies, a human would need to drink dozens of bottles of wine a day. Resveratrol sparked a frenzy, but in time, the enthusiasm sobered.

Still, the core insight remained. Sirtuins are real, and powerful. They integrate energy status, environmental stress, and aging. While resveratrol wasn't the miracle many hoped, it lit the path toward compounds—and behaviors—that could activate these ancient pathways.

Today, these insights have filtered into everyday routines. You've likely heard of them: 16:8, OMAD, alternate-day fasting. These aren't just diet trends—they're modern echoes of ancient biology.

In a 16:8 fast, for instance, you eat all your calories in an 8-hour window, then go 16 hours without food. That's enough to lower insulin, reduce IGF-1, and activate AMPK and sirtuins. In lab animals, this timing alone—without changing what they eat—extends lifespan and improves markers of health.

Fasting mimics the evolutionary stressors our bodies were designed to withstand. It's not about deprivation—it's about rhythm. Feeding tells cells to grow. Fasting tells them to repair.

Some studies even suggest that brief fasts enhance cognitive function. Mice deprived of food for short periods show increased brain-derived neurotrophic factor (BDNF), which supports learning and memory. Hunger, in this light, becomes a kind of signal flare for the brain: pay attention, adapt, survive.

The desert hunter knew this without knowing it. His sharp senses, alert movements, and clarity of mind weren't accidental—they were adaptive responses to fasting. His body wasn't failing; it was thriving under pressure.

Of course, fasting is not a one-size-fits-all solution. There are risks, especially for people with eating disorders, diabetes, or certain medical conditions. And long-term, extreme restriction can backfire—causing muscle loss, hormonal imbalance, and psychological stress.

But the contrarian insight still stands: fasting isn't about suffering or shrinking your waistline. It's about biochemical choreography—switching on the genes and pathways that restore balance, eliminate damage, and extend functional life.

For most of modern history, the emphasis has been on what to eat: carbs or fat, meat or plants, keto or vegan. But fasting flips the frame. It's not just what we eat—it's when we stop eating—that matters.

In silence, the body listens. In stillness, it repairs.

LESSONS FROM LAB RATS

In a quiet facility tucked behind a university biology department, rows of rodents live out what might be the cleanest, calmest lives any mouse could hope for. Temperaturecontrolled rooms. Filtered air. A perfectly measured, precisely timed diet—not a crumb more than their daily ration. They are not starving. But they are not full either.

To a visitor, the calorierestricted mice are uncanny. Sleek coats, brighter eyes, unusual agility. In a species that typically declines into frailty by two years of age, these mice still scurry with youthful enthusiasm. They seem like time travelers—aging, but not deteriorating. Dying, eventually, but not collapsing.

This is the strange magic of caloric restriction, or CR: a reduction in food intake without malnutrition that dramatically extends lifespan in a wide variety of species. In some cases, up to 50% longer. The data isn't subtle. It's robust, reproducible, and surprisingly old. The first study was published in 1935. Nearly a century later, the core findings remain unchanged.

Yeast cells, when deprived of glucose, live longer. Worms, flies, and fish all show extended lifespans under reduced caloric intake. Mice, the most frequently studied mammals, respond in kind: lower food, longer life. The effect is so consistent that CR is often called the gold standard of anti-aging interventions in lab models.

But the leap from mice to men is treacherous. What works for a mouse may not work for a monkey—or a middle-aged human with a mortgage and a sweet tooth.

To test this, researchers turned to our closest cousins: rhesus monkeys. In two major studies—one at the National Institute on Aging (NIA), the other at the University of Wisconsin—monkeys were placed on 30% calorierestricted diets for decades. The results were striking, but complicated.

In the Wisconsin study, calorierestricted monkeys looked visibly younger. They had less muscle loss, fewer chronic diseases, and lower incidence of cancer and diabetes. Some lived longer. But in the NIA study, although CR monkeys were healthier, they didn't live significantly longer than controls.

The takeaway? CR improves healthspan—the period of life spent free from disease—even if it doesn't always extend lifespan in every context. In humans, where causes of death vary wildly and social complexity muddies every variable, the picture gets even murkier.

Still, the idea was too tantalizing to ignore.

In the early 2000s, a dedicated group of humans began applying CR to themselves. Enter the Calorie Restriction Society—an organization of people voluntarily reducing their food intake by 20—30% in the hope of extending their lives. They weigh their spinach, skip dessert, log every calorie, and proudly sport resting heart rates in the 40s. Some are in their 50s but boast the cholesterol profiles of teenagers. Their commitment borders on monastic.

A few of them allowed researchers to study their physiology. The results were, again, impressive. Participants showed lower inflammation, reduced blood pressure, better insulin sensitivity, and dramatically improved lipid profiles. By the numbers, they were aging more slowly.

Internally, their bodies behaved younger than their chronological age.

But the benefits came with costs.

Some CR practitioners experienced bone density loss—a possible consequence of chronically low energy intake. Others reported decreased libido, mood instability, and difficulties maintaining muscle mass. For women, extended caloric restriction sometimes disrupted menstrual cycles. Evolution had prepared the human body for short bouts of hunger—not permanent semi-starvation.

And then there was the mental toll. One long-term practitioner admitted: "I feel great most days. But some nights, when my family is eating cake and I'm sipping herbal tea, I wonder if I've traded joy for metrics."

That tension—between quantity of life and quality of living—hangs over the entire CR movement. There is no denying the biological benefits. But there is also no escaping the tradeoff.

Perhaps the better question is: what do we want from longevity?

In mice, CR works because it flips metabolic switches—reducing mTOR and IGF-1, increasing AMPK and sirtuins, ramping up autophagy and stress resistance. These are real mechanisms, measurable and consistent. But humans are more than mechanisms.

The promise of CR is not just about living longer—it's about delaying decline. Moving into old age with less pain, more function, more freedom. That doesn't always mean eating 30% less. It might mean timing meals differently, eating with intention, giving the body periods of rest between feasts.

Moderation may not be sexy, but it may be sustainable. What CR teaches us is not that we should all shrink our plates to mouse-sized portions. It's that our bodies are sensitive to signals, and that less can be more—but only up to a point. Beyond that, restriction becomes stress. And not the kind that builds resilience.

Back in the lab, the CR mice continue their measured lives. They will die later than their siblings—but not as martyrs. Just as participants in an experiment that began before most of us were born.

THE LONGEVITY TOGGLE

In a quiet corner of Okinawa, not far from the sea, a woman in her early hundreds bends over a small vegetable patch. Her movements are slow but not frail. She weeds with bare hands, wipes her brow with a cloth tucked into her apron, then stands to admire the ridges of purple sweet potato vines snaking toward the sunlight. A neighbor calls out to her. She smiles. Later, she'll prepare a simple lunch—miso soup, a sliver of grilled fish, maybe a handful of goya. No dessert. No calorie tracker. No protein shake. Just food. Just enough.

She has never heard of mTOR, or AMPK, or the insulin/IGF axis. But her life, lived in rhythm with hunger, laughter, and light, has managed to toggle the molecular switches that modern medicine is only beginning to understand.

Aging, at its molecular root, is a matter of balance disturbed. In youth, the body is an empire of growth—cells divide, tissues build, reproduction thrives. Growth signals like insulin, IGF-1, and mTOR shout their instructions across the cytoplasmic airwaves: build, expand, multiply. But as we age, this empire refuses to yield power. Growth signals continue to roar even when no one is left to listen. Repair becomes an afterthought. Cleanup stalls. The system exhausts itself under the weight of unchecked abundance.

This is the essence of deregulated nutrient sensing—a hallmark of aging. But the human body, ever adaptable, has built-in toggles to shift between modes: from growth to conservation, from feast to fast. This toggle is not binary. It is fluid, subtle, rhythmic. And in many of the longestlived people on Earth, it still works—not because of genetic privilege, but because of how they live.

The scientific community has begun to probe these switches with increasing sophistication. Nutrient cycling—the alternation between states of abundance and scarcity—is emerging as one of the most promising frameworks for extending both lifespan and healthspan. It turns out that the secret may not lie in constant restriction, but in strategic fluctuation. Like seasons, the body thrives when there are cycles: of growth, of rest, of cleanup.

Take rapamycin, a compound discovered in the soil of Easter Island. It inhibits mTOR, the master controller of cellular growth. In mice, intermittent dosing of rapamycin—even when started late in life—has been shown to extend lifespan and reduce age-related decline. It's not about shutting down growth permanently, but about reminding the body when to stop building and start fixing.

Or metformin, the common diabetes drug. It activates AMPK, the cellular fuel gauge that kicks in when energy is low. AMPK promotes glucose uptake, mitochondrial health, and autophagy—the internal recycling process that clears damaged components. In large observational studies, diabetics on metformin have outlived their non-diabetic peers. A drug meant for blood sugar control is now being studied for its anti-aging potential.

The there's spermidine, a molecule found in fermented foods and aged cheese, which appears to stimulate autophagy as well. Early studies suggest it may help maintain cardiovascular and cognitive function with age. Together, these compounds form the pharmacological vanguard of metabolic rebalancing—restoring the body's ability to shift gears when needed.

But the truth is, nature invented this toggle long before we bottled it.

Every time we fast, even briefly, the body lowers insulin, quiets mTOR, and activates cleanup crews like AMPK and sirtuins. Every time we exercise, we trigger a burst of oxidative stress that signals the body to adapt and grow stronger. This is hormesis—the idea that small, controlled doses of stress build resilience. It's why heat exposure (like saunas) and cold exposure (like ice baths) are being explored as longevity tools. The stress is not the enemy. It is the stimulus for renewal.

But the message, once again, is moderation. Chronic starvation, obsessive overtraining, or unrelenting stress do not produce longevity—they erode it. The goal is not to silence growth permanently, but to restore the body's capacity to balance growth with repair. In modern societies saturated with food, comfort, and stimulation, the toggle is stuck in one position. The remedy isn't suffering—it's metabolic wisdom.

And that wisdom often lives in the most unassuming places.

Reader-tool boxes for this chapter

CLOSER LOOK ***Kenyon's worms — when a single gene doubled lifespan***

In 1993, Cynthia Kenyon's lab at UCSF reported that adult Caenorhabditis elegans hermaphrodites carrying a mutation in the daf-2 gene lived more than twice as long as wildtype worms — and stayed active and fertile while they did. The mutation needed a second gene, daf-16, intact to deliver its effect. Kenyon's group had stumbled into the worm version of the insulin/IGF-1 signaling pathway.

The result detonated a long-standing assumption: that aging was not genetically programmed and not modifiable by single mutations. If one gene could double lifespan in a worm without obvious cost, the field had to take seriously the idea that aging was regulated by signaling pathways

— pathways that could be drugged. Every modern longevity trial, from rapamycin to metformin, traces its intellectual lineage back to those long-lived worms.

Kenyon, Chang, Gensch, Rudner, & Tabtiang (1993). A C. elegans mutant that lives twice as long as wild type. Nature, 366(6454), 461–464.

EASY ACCESS ***mTOR vs. AMPK in plain English***

Your cells have two opposing voices that decide how to behave: a feast voice and a famine voice. mTOR is the feast voice — when amino acids and insulin are high, it tells the cell to build, divide, and store. AMPK is the famine voice — when energy gets low, it tells the cell to conserve, repair, and clean house. In youth the two voices alternate. With chronic overeating and chronic insulin signaling, the feast voice never gets quiet, the cleanup voice rarely gets a turn, and the cell starts to drown in its own neglected garbage. Most longevity interventions — fasting, exercise, rapamycin, metformin — work by giving AMPK the floor.

CLOSER LOOK ***Rapamycin — the first true mammalian gerotherapeutic***

In 2009, the NIA's Interventions Testing Program reported the first replicated demonstration of a drug extending lifespan in a mammal. Rapamycin, fed to genetically heterogeneous mice starting at 600 days of age — the rodent equivalent of beginning treatment in late middle age — extended median and maximal lifespan in both sexes. The effect held across three independent labs.

Rapamycin works by inhibiting mTORC1, the master signaling complex that integrates amino acid availability, insulin, and growth signals. Subsequent work showed lifespan extension when treatment began even later in life, and parallel studies in dogs (the Dog Aging Project) are now testing whether the same biology translates to companion animals. Use in healthy humans remains experimental — the drug carries real immunosuppressive risks at chronic doses — but as proof of principle, it broke the field open.

Harrison et al. (2009). Rapamycin fed late in life extends lifespan in genetically heterogeneous mice. Nature, 460(7253), 392–395.

CLOSER LOOK ***Caloric restriction in monkeys — a controversy with a resolution***

Two long-running studies tested whether caloric restriction extends life in rhesus macaques. The University of Wisconsin-Madison study reported a clear survival benefit. The National Institute on Aging study, run in parallel, reported essentially none. For years the field treated this as a contradiction.

A 2017 joint reanalysis of both cohorts resolved the puzzle. The studies differed in baseline diet quality, age at intervention onset, and the genetic background of the monkeys. When those factors were accounted for, the data converged: caloric restriction without malnutrition does extend healthspan and survival in nonhuman primates, but the magnitude depends on what the comparison diet looks like and when in life the restriction begins. The clean lesson — eat less to live longer — turned out to need three asterisks. The deeper lesson, that nutrient signaling shapes the aging trajectory, held up.

Mattison et al. (2017). Caloric restriction improves health and survival of rhesus monkeys. Nature Communications, 8, 14063.

EASY ACCESS *Metformin and the TAME trial — what's actually being tested*

Metformin is cheap, generic, and has been used for type 2 diabetes since the 1950s. Observational studies kept turning up something odd: diabetics on metformin had lower rates of cancer, lower rates of cardiovascular events, and sometimes longer lifespans than non-diabetics in the same age band. Mechanistically, the drug activates AMPK and dampens mTOR signaling — the same axis we just discussed. The TAME trial (Targeting Aging with Metformin), led by Nir Barzilai at Einstein, is designed to test whether metformin, given to non-diabetic older adults, delays the onset of multiple age-related diseases. It is not a longevity trial in the colloquial sense. It is a regulatory experiment — testing whether the FDA will recognize aging itself as an indication.

KEY TERMS *The vocabulary of the nutrient sensor*

AMPK — AMP-activated protein kinase. The cell's low-energy alarm; activates autophagy and shuts down anabolism.

mTOR — Mechanistic target of rapamycin. Master signaling complex that fires when nutrients are abundant, driving growth and suppressing autophagy.

mTORC1 (mechanistic target of rapamycin complex 1) — The cell's master growth switch. When active, it tells the cell to build proteins, divide, and grow; when quiet, it allows the cell to recycle and repair. Chronically active mTORC1 is one of the most consistent signatures of accelerated aging.

Leucine — A branched-chain amino acid found in high concentrations in meat, dairy, eggs, and whey protein. The most potent single dietary signal that flips mTORC1 from off to on.

Methionine — An essential amino acid concentrated in meat, fish, eggs, and dairy; lower in most plant proteins. Restricting methionine alone, without cutting total calories, extends maximum lifespan in rodents.

Sestrin2 — A cytosolic protein that binds leucine directly and serves as the leucine sensor for mTORC1. Empty-handed, Sestrin2 holds the upstream complex (GATOR2) inactive; leucine binding releases it and lets mTORC1 fire.

GATOR1 and GATOR2 — A pair of opposing complexes at the lysosome that decide whether mTORC1 turns on. GATOR1 keeps mTORC1 off; GATOR2 blocks GATOR1. Sestrin2 inhibits GATOR2 — so when leucine binds Sestrin2 and frees GATOR2, the brake on the brake is lifted, and mTORC1 fires.

IGF-1 (insulin-like growth factor 1) — A growth-promoting hormone released largely by the liver in response to dietary protein. Persistently elevated IGF-1 is linked to higher cancer mortality in middle age; mice and humans with reduced IGF-1 signaling tend to live longer and develop fewer age-related diseases.

Insulin/IGF-1 signaling — The pathway downstream of insulin and insulin-like growth factor 1. Strongly nutrient-responsive; tightly linked to lifespan across species.

Branched-chain amino acids (BCAAs) — Leucine, isoleucine, and valine. Marketed heavily for muscle growth, which they do support. The longevity question is whether chronic high intake — through supplements or animal-heavy diets — keeps mTORC1 turned up past the point of benefit.

Rapamycin — An mTOR inhibitor. The first drug shown to extend lifespan in a mammal.

Sirtuins — A family of NAD+-dependent enzymes (SIRT1–7) that respond to metabolic state and regulate stress, repair, and metabolism.

Caloric restriction — Reduction in energy intake without malnutrition. The longest-studied lifespan-extending intervention in animals.

POWERHOUSE ON THE FRITZ

MITOCHONDRIAL DYSFUNCTION

He'd done this hike every summer since he was fourtytwo. From the weathered wooden gte at the trailhead to the lookout over Lac Bleu, it was just under five miles—mostly forest path, one steep switchback, a bit of scree near the top. Every bend had a memory: where he once stopped to tie his daughter's bootlace, where the ibex stood unmoving one spring, where he dropped his flask and let it go.

But this time was different.

The air was crisp, the sun indifferent, the same pinesweet scent blanketing the path—but his legs betrayed him. Before the halfway mark, his calves began to burn. At the incline, his breath quickened unnaturally, and by the last slope, he had to sit. Not from pain, but exhaustion—deep, cellular fatigue. He tried to laugh it off to the couple passing him—"Must've brought the wrong legs"—but it wasn't funny. It was unfamiliar. It was frightening.

Later that week, seated stiffly in a Zurich sports medicine clinic with more ambition than decoration, he described the sensation to a young doctor with a tablet and a practiced smile.

"It's like the power's out," he said.

She paused, eyes flicking to his chart. "That's actually a good way to put it."

She showed him a diagram. Not of muscles, or bones—but of mitochondria. "These," she said, "are your cells' batteries. Tiny power plants. And as we age, they begin to falter."

Mitochondria have long been overlooked in high school biology class—reduced to their punchline nickname as "the powerhouse of the cell." But that phrase, bland as it sounds, is literally true. Mitochondria are the energy engines of life. Inside each of your trillions of cells, they churn out adenosine triphosphate (ATP)—the energy currency that drives everything from muscle contraction to brain signaling to immune surveillance. Without ATP, nothing moves, nothing thinks, nothing repairs.

KEY TERMS *The vocabulary of the powerhouse*

Mitochondrion — A small organelle inside almost every cell whose main job is to produce ATP. Densely packed in muscle, brain, and heart.

ATP (adenosine triphosphate) — The energy currency of the cell. Made in mitochondria, spent everywhere else — by muscles, neurons, immune cells, and the machinery of repair.

ROS (reactive oxygen species) — Highly reactive oxygen-containing molecules. Byproducts of mitochondrial energy production. In small doses, signals; in excess, damage.

Oxidative phosphorylation — The process mitochondria use to convert oxygen and food into ATP.

Up to fifteen times more efficient than energy production without oxygen.

Mitochondrial DNA (mtDNA) — A small, circular genome carried inside every mitochondrion — a remnant of an ancient bacterial ancestor. Inherited only from your mother.

Mitophagy — The cell's quality-control system for mitochondria. Damaged mitochondria are tagged, enclosed, and digested before they can pollute the cell.

Mitochondrial biogenesis — The birth of new mitochondria from existing ones. Triggered by exercise, fasting, and cold.

What makes mitochondria special isn't just their function, but their origin. These tiny organelles are thought to be the descendants of ancient bacteria that, more than a billion years ago, formed a symbiotic relationship with primitive cells. The endosymbiotic theory suggests that one cell swallowed another, but instead of digesting it, they struck a deal: protection in exchange for power. Over time, that engulfed bacterium became mitochondria. Today, they still carry their own DNA—small, circular, and maternally inherited—a ghost of their independence.

The arrangement worked brilliantly for millennia. Mitochondria evolved to be efficient ATP factories, using oxygen to break down glucose and fatty acids through a process called oxidative phosphorylation. The payoff is tremendous: this process generates up to 15 times more ATP than anaerobic metabolism. That's why the brain and muscles—both energyhungry tissues—are densely packed with mitochondria.

But there's a catch. Like any high-efficiency system, mitochondria also produce waste: reactive oxygen species (ROS). These highly reactive molecules are like sparks from a combustion engine—tiny but potentially destructive. In youth, cells neutralize ROS with antioxidant defenses. But as we age, this balance tips. Mitochondria become less efficient. They leak more ROS. Their DNA, located right next to the combustion zone, gets damaged. And because mitochondrial DNA lacks the robust repair mechanisms of nuclear DNA, those mutations accumulate.

> **EASY ACCESS** ***What's actually happening when you feel the "power's out"***
>
> Your cells run on a tiny battery called ATP. Mitochondria make it. As you age, those batteries get smaller, leak more, and replace themselves more slowly. Less ATP per breath means muscles tire faster, brains fog earlier, and recovery takes longer. The mountaineer wasn't losing his fitness — he was running short on the molecule that made fitness possible.

The result? A slow but insidious decline. Cells get less energy, more damage. Tissues start to underperform. In the mountaineer's case, that meant burning muscles and short breath. In others, it might be forgetfulness, fatigue, or poor healing. On a cellular level, it's the power flickering. On an organismal level, it's aging.

The mountaineer stared at the screen as the doctor explained mitochondrial dysfunction. "It's not that your cells are broken, " she said, " but their energy output is reduced. They're slower to respond. They produce more exhaust. Over time, it adds up."

He asked what he could do. She smiled again—less rehearsed this time. "That's the good news," she said. "There's still a lot we can do to keep them running longer."

This is where our journey into mitochondria begins. With people and their fading stamina. With organelles that are both ancestors and engines. With a theory that started with radiation and ended with the realization that aging is not just decay—it's dysfunction. And sometimes, dysfunction begins at the power source.

In the next part, we'll trace how this dysfunction doesn't just sap energy, but contributes to a cascade of biological consequences—from cellular senescence to systemic inflammation—and why the idea that "more antioxidants = more youth" may be far too simple.

But first, it helps to understand that the blackouts aren't just happening in the mountains. They're happening inside us. Quietly. Relentlessly. And like all infrastructure failures, they begin long before the lights go out.

THE OXIDATIVE MIRAGE

In the fall of 1956, a man sat hunched in a dimly lit office in Nebraska, rifling through scientific journals not meant for his field. Denham Harman was a chemist by training, not a gerontologist. Yet something in the literature on radiation damage caught his attention. Radiation, he knew, created chaos in cells by unleashing molecules that shredded DNA, mangled proteins, and sabotaged cellular machinery. What if the body, even in the absence of radiation, made its own version of these destructive agents? Molecules that invisibly chipped away at us, one cell at a time?

That flash of insight gave birth to what would become one of the most influential—and later controversial—theories of aging: the Free Radical Theory. Harman proposed that our cells, in the very act of making energy, also generate tiny saboteurs known as reactive oxygen species (ROS). These were not foreign invaders, but the price of being alive. To understand how audacious this was, consider the dual nature of mitochondria. These microscopic organelles, nestled in nearly every cell in the body, operate like coalfired power plants. Their job is to take fuel—glucose, fat—and convert it into ATP, the energy currency of life. But just as real power plants emit smoke and soot, mitochondria release byproducts: free radicals. These molecules have unpaired electrons, making them highly unstable and reactive. In small amounts, they're a normal part of metabolism. In excess, they become biochemical wrecking balls.

CLOSER LOOK ***Where the sparks come from — the electron transport chain***

Inside the inner membrane of every mitochondrion sits a series of four protein complexes — the electron transport chain. As food is broken down, electrons enter Complex I and pass, like buckets in a fire line, from complex to complex. At each handoff, a small amount of energy is harvested to pump protons across the membrane. The proton gradient drives ATP synthase, the molecular turbine that makes ATP. The system is staggeringly efficient — but not perfect. Roughly one to two percent of the electrons "leak," colliding with oxygen to form superoxide, the parent reactive oxygen species. With age, the complexes themselves degrade, and the leak rate climbs. The same machinery that powers life produces, as a byproduct, the molecules that wear it down.

Harman, D. (1956). Aging: A theory based on free radical and radiation chemistry. Journal of Gerontology, 11(3), 298–300.

In youth, the body manages this beautifully. Cells deploy antioxidant enzymes like glutathione peroxidase and catalase—molecular janitors that patrol and neutralize excess ROS. It's an elegant balancing act: energy generation on one side, cleanup on the other.

But with age, the choreography falters. The balance tilts. Studies across tissues—from liver slices to human neurons—show that aging cells produce more ROS while simultaneously mounting a weaker antioxidant response. This is not a subtle shift. Oxidative stress levels in old cells can be severalfold higher than in young ones. The consequences ripple outward:

proteins misfold, lipids in membranes turn rancid, DNA strands fracture. These aren't cosmetic dings. They change how cells function—or if they survive at all.

Perhaps most crucially, the very structure tasked with making ROS—the mitochondrion—starts to devour itself. Mitochondrial DNA (mtDNA), which exists in close proximity to the electron transport chain where ROS is generated, lacks the protective histones and robust repair mechanisms that shield nuclear DNA. The result? Mutations accumulate. Damaged mtDNA leads to faulty mitochondrial proteins, which in turn create more ROS in a vicious feedback loop. It's the biochemical equivalent of an engine throwing a rod and then flooding the car with smoke.

CLOSER LOOK ***How a cell decides to scrap a mitochondrion***

Healthy mitochondria maintain a voltage across their inner membrane — the spark, in molecular terms, that makes ATP possible. When a mitochondrion is damaged enough to lose that voltage, an enzyme called PINK1 stops being imported and degraded inside it, and instead accumulates on the outer membrane. PINK1 is a beacon. It recruits a second protein, Parkin, which begins decorating the failed mitochondrion with ubiquitin — the cellular equivalent of a "demolish me" tag. Tagged mitochondria are engulfed by a double-membraned vesicle called an autophagosome and trafficked to the lysosome, where they are dismantled and their parts recycled. With age, expression of PINK1 and Parkin drops, lysosomal function wanes, and the whole pipeline slows. Damaged mitochondria linger. The cell ends up running on a fleet of aging, leaky engines it cannot afford to replace.

Pickrell & Youle (2015). The roles of PINK1, parkin, and mitochondrial fidelity in Parkinson's disease. Neuron, 85(2), 257–273.

This self-reinforcing cycle helps explain why oxidative stress is not just a footnote in aging but a central narrative. It links directly to major age-related diseases. In Alzheimer's, oxidative stress precedes and possibly drives the formation of amyloid plaques. In cardiovascular disease, oxidized LDL cholesterol is a known contributor to arterial plaque. Even in type 2 diabetes, ROS interfere with insulin signaling and betacell survival.

And yet, the story doesn't end here.

For decades, the logical conclusion seemed obvious: more ROS means more aging, so let's flood the system with antioxidants. This spawned a billiondollar supplement industry and decades of clinical trials. People popped vitamin C, E, and betacarotene pills like life-extending candy. But something strange happened. The trials didn't pan out. In fact, in some studies, high-dose antioxidants slightly increased mortality. The smoke didn't clear—it thickened.

What went wrong?

The problem, as it turns out, was not with Harman's observation but with our interpretation. Oxidative stress is real. The damage is real. But the idea that all ROS are harmful was too simplistic. It missed the subtler truth: that ROS are also signals. Like sparks from a fire, they can burn—but they can also light the way.

Cells use small bursts of ROS to activate defense pathways. These include stress response genes, autophagy, and even anti-inflammatory signals. Wipe out ROS entirely, and the body may never hear the alarm bells. In fact, one study showed that mice genetically engineered to produce extra antioxidant enzymes didn't live longer. Another found that athletes who took antioxidant supplements had fewer fitness gains—because the ROS that exercise normally triggers are part of how muscles adapt and grow stronger.

CLOSER LOOK ***When the cure made things worse — what the antioxidant trials actually showed***

The most rigorous test of the antioxidant hypothesis came in 2007. Bjelakovic and colleagues pooled data from 68 randomized trials covering more than 230,000 participants. When the analysis was restricted to the 47 trials with the lowest risk of bias, the result was the opposite of what the supplement industry had promised. Beta-carotene, vitamin A, and vitamin E — singly or combined — were associated with a small but statistically significant increase in all-cause mortality. The reasons are still debated, but a key piece of the puzzle came from a 2008 trial of vitamin C in trained men. The supplemented group built fewer mitochondria in response to endurance training. The vitamin C had blocked the very ROS signal that tells muscle to adapt. Antioxidants, in megadose form, interrupt the conversation between stress and resilience.

Bjelakovic et al. (2007). Mortality in randomized trials of antioxidant supplements. JAMA, 297(8), 842–857; Gomez-Cabrera et al. (2008). American Journal of Clinical Nutrition, 87(1), 142–149.

What Harman ignited in 1956 was not a false theory, but an incomplete one. Oxidative stress is a layer of the aging puzzle, not the whole picture. ROS are both villains and messengers, depending on context, dose, and timing. The mitochondrion, far from being a mindless generator of destruction, is more like a monk with a forge: producing both power and lessons, suffering damage but also sending signals for healing.

RECYCLING GONE WRONG

The old waste management plant on the edge of Kyoto hadn't been upgraded since the late 1970s. For decades, it served the city well—sorting plastics, metals, and biodegradables with nearreligious precision. But something changed in recent years. The workers grew fewer, the machinery slower. Soon, recyclables began slipping into incinerators, toxic fumes rising over cherry blossoms. Eventually, trucks stopped unloading altogether. Containers piled up, and neighborhoods began to notice the stench. What once ran like a clean, closedloop system now hemorrhaged waste into the streets.

Aging cells are no different when it comes to managing waste—especially the wornout engines of life: mitochondria. In youth, cells constantly monitor mitochondrial performance, tagging the underperformers for removal. This system is known as mitophagy, a specific form of autophagy—think of it as quality control for cellular power plants. When it works, it's invisible. When it doesn't, the garbage leaks into the rest of the cell, poisoning everything in sight.

Mitophagy is initiated when a mitochondrion becomes depolarized—essentially, when it loses its membrane potential and can no longer generate ATP effectively. Under these conditions, an enzyme called PINK1, normally degraded in healthy mitochondria, starts to accumulate on the outer membrane. That's the signal for another protein, Parkin, to swing into action. Parkin tags the failing mitochondrion with a molecular "remove me" label, prompting its enclosure by a doublemembraned vesicle and eventual digestion in the cell's lysosomal stomach. It's a clean kill. Efficient. Essential.

But like Kyoto's aging waste plant, this process degrades over time.

Studies in aged animals and humans show a striking decline in mitophagy. One paper, published in Nature Communications, demonstrated that muscle cells from elderly individuals had significantly lower levels of PINK1 and Parkin activity. These cells weren't replacing their defective mitochondria—they were accumulating them. The researchers likened it to driving with flat tires and refusing to change them. You can still move forward, technically—but the wear and inefficiency mount quickly.

What happens when cells hoard broken mitochondria? The consequences are both subtle and systemic. First, damaged mitochondria generate excess reactive oxygen species (ROS), which, as we saw earlier, harm proteins, lipids, and DNA. But second—and perhaps more dangerously—these broken mitochondria begin to leak fragments of their own DNA into the cytosol. This mitochondrial DNA (mtDNA), while technically self-derived, shares an evolutionary resemblance to bacterial genomes. To the immune system, it looks like a foreign invader.

This sets off a chain reaction. Innate immune receptors like the cGAS-STING pathway detect the mtDNA and launch an inflammatory response. Cytokines flood the cellular environment, triggering nearby cells to enter defense mode. If the leakage continues, this localized response can escalate into systemic inflammation. The result is what scientists now call inflammaging—a chronic, low-grade inflammatory state that becomes more pronounced with age and is a hallmark of virtually every age-related disease, from Alzheimer's to atherosclerosis.

EASY ACCESS ***Why a tired mitochondrion makes the whole body inflamed***

Mitochondria descend from ancient bacteria, and their DNA still looks bacterial. As long as it stays inside the mitochondrion, the immune system ignores it. But when an old, leaky mitochondrion spills that DNA into the cell, the immune system reacts as if a bacterial infection has begun — and starts firing alarms. Multiply this across millions of cells and decades, and you get the slow, smoldering inflammation that drives heart disease,

Alzheimer's, and frailty. Your own batteries, leaking, are mistaken for invaders.

And the waste is not inert. Damaged mitochondria are active propagators of dysfunction. Their ROS signals don't just harm—they misinform. They can push neighboring cells into senescence, where those cells stop dividing but begin secreting their own cocktail of proinflammatory chemicals, known as the senescence-associated secretory phenotype (SASP). It's a biochemical whisper campaign gone rogue. Over time, the cellular environment becomes hostile. Repair slows. Regeneration halts. Tissue integrity declines.

In muscles, this manifests as frailty—a creeping loss of strength, balance, and endurance.But the consequences aren't limited to muscle. In the immune system, failing mitophagy allows aged immune cells—particularly macrophages and neutrophils—to become sluggish and misfire. They release inflammatory signals even in the absence of threats, attacking tissues that don't need defending. This helps explain why autoimmune disorders rise with age, and why older individuals are more vulnerable to complications from infections like COVID-19. Their immune systems are operating on outdated, inaccurate energy maps.

Why does mitophagy decline with age? The causes are multifactorial. Expression of PINK1 and Parkin drops. Lysosomal function—the final stage of digestion—wanes. The sensors that detect mitochondrial dysfunction lose sensitivity. Like the Kyoto plant, it's not one broken cog but many, interacting over time. And because mitophagy is not optional—because failing to remove toxic mitochondria affects every tissue—it becomes a silent but central culprit in the aging process.

There's reason for hope, however. Interventions that restore mitophagy show promise in animal models. Endurance exercise remains the gold standard: it stimulates mitochondrial biogenesis and appears to upregulate mitophagy, even in aged tissues. New molecules like urolithin A—produced by gut bacteria from ellagitannins found in pomegranates—have been shown to induce mitophagy in muscle cells and improve strength in older adults. Researchers are also exploring compounds that directly activate PINK1/Parkin signaling or enhance lysosomal function.

CLOSER LOOK *A microbe-made molecule that tells old muscle to clean house*

When you eat a pomegranate, certain gut bacteria in some people convert compounds called ellagitannins into a small molecule named urolithin A. In a 2019 first-in-human trial, healthy older adults given urolithin A showed transcriptional signatures in muscle biopsies consistent with activated mitophagy — the same housecleaning process that wanes with age. Subsequent trials have reported modest gains in muscle endurance and improvements in mitochondrial gene expression. None of this is a fountain of youth, and long-term data are still being gathered. But the underlying principle is the important part: an exogenous molecule can nudge a cell to clean up its own mitochondria, and old cells appear willing to listen.

Andreux et al. (2019). The mitophagy activator urolithin A is safe and induces a molecular signature of improved mitochondrial and cellular health in humans. Nature Metabolism, 1(6), 595–603.

Still, there's no magic bullet. You can't take a pill and expect decades of cellular junk to vanish overnight. But the broader insight—that aging can be slowed by restoring the body's internal recycling system—is profound. It tells us that the problem isn't just decay. It's mismanagement.

TOO MUCH OF A GOOD THING

The 1990s were peak antioxidant mania. Bottles of vitamin E, betacarotene, and coenzyme Q10 flew off pharmacy shelves. Health columns declared war on free radicals. Supermarkets were suddenly flooded with "antioxidant-rich" labels: blueberries, green tea, dark chocolate. It was an era of optimism dressed in capsules. If oxidative stress accelerated aging, then neutralizing free radicals must be the answer. The logic was seductive. But science rarely obeys tidy syllogisms.

By the 2000s, the shine began to fade. A large metaanalysis published in the Journal of the American Medical Association examined data from nearly 70 trials involving over 230,000 people. The results were jarring: not only did antioxidant supplements fail to reduce mortality—they slightly increased it in some groups. Vitamin A and E, once thought to be life extenders, were now associated with greater risk of early death.

It wasn't a quiet fizzle. It was a fullon contradiction. How had the miracle cure turned toxic?

To understand this paradox, we have to reexamine the role of reactive oxygen species (ROS) through a more sophisticated lens—one that recognizes them not just as villains, but as reluctant messengers.

ROS are byproducts of metabolism, especially in mitochondria. Like sparks from a fire, they can ignite damage when uncontrolled. But in measured doses, they also signal cells to adapt, strengthen, and defend. This dual role—a poison and a prompt—is the foundation of a concept called mitohormesis. Hormesis, broadly, is the biological principle that a small amount of stress can be beneficial. Vaccines use this logic. So does physical training. The body, challenged mildly, responds with resilience.

Mitochondria play a central role in mitohormesis. When a person exercises, their muscles consume more oxygen. Mitochondria rev up, and ROS levels temporarily spike. Rather than causing harm, this controlled oxidative stress flips genetic switches that activate antioxidant defenses, DNA repair, and the creation of new mitochondria. It's as if the cell reads the stress as a rehearsal for harsher times and bolsters its arsenal. This, in part, explains why exercise is one of the most potent anti-aging interventions known—not because it reduces ROS, but because it induces them wisely.

CLOSER LOOK *Mitohormesis — when a little stress is the medicine*

When a person exercises, oxygen consumption in working muscles can rise by a factor of fifty. Mitochondria respond by producing a brief surge of ROS. Those ROS, rather than causing damage, activate a cascade of stress-response genes — including PGC-1α, the master regulator of mitochondrial biogenesis, and Nrf2, the transcription factor that upregulates the body's endogenous antioxidant enzymes. The cell builds more mitochondria and more antioxidant capacity. Ristow and colleagues demonstrated this directly: subjects who supplemented with high-dose vitamin C and E during a four-week training program saw blunted improvements in insulin sensitivity and reduced expression of antioxidant genes. The supplements had erased the very signal that exercise was trying to send. The lesson: stress, in measured doses, is information. Suppress the signal, and you suppress the adaptation.

Ristow et al. (2009). Antioxidants prevent health-promoting effects of physical exercise in humans. Proceedings of the National Academy of Sciences, 106(21), 8665–8670.

Here's where it gets more complicated. In a landmark

study, scientists tested whether high-dose antioxidant supplements would enhance or undermine this adaptive benefit. One trial involved two groups of men performing regular endurance training. One group received no supplements. The other received 1000 mg of vitamin C daily. The results? The supplement group showed less improvement in insulin sensitivity and mitochondrial biogenesis. Essentially, the antioxidants had blocked the very signals that prompted the body to adapt. They short-circuited the stress.

This was not a fluke. Other studies confirmed similar findings: athletes who supplemented heavily with antioxidants saw reduced gains from training. It was a biochemical betrayal—attempting to help, the pills had dulled the body's internal feedback loop.

The idea that all oxidative stress is bad now seems laughably outdated. A more nuanced view is taking hold. Like the immune system, the cell's stress responses require calibration, not suppression. Just as an overactive immune system can cause autoimmune disease, an underactive one leaves us vulnerable. Similarly, if we douse every molecular spark, we never allow the body to learn from the flame.

This also explains why mice engineered to produce excess antioxidant enzymes don't necessarily live longer. In some models, they aged just like controls. In others, they fared worse. The body, it turns out, evolved to navigate an oxidative environment. ROS serve as traffic signals, not just debris. When stripped away entirely, those signals vanish.

Even in simple organisms, low-level ROS can be beneficial. In one famous experiment, C. elegans worms were exposed to paraquat—a pesticide that increases ROS production. Instead of dying young, the worms lived longer. The ROS had triggered their defense programs, turning on a suite of genes involved in stress resistance. The worms adapted. They braced themselves and survived better. The dose, of course, was critical—too much paraquat and they died. But the lesson stood: the danger wasn't in ROS alone, but in misunderstanding their role.

One might ask: If antioxidants are still essential (and they are), where should we get them? The answer is less controversial now: from food, not pills. Whole foods offer a complex blend of antioxidants, fiber, enzymes, and phytochemicals that interact with our biology in ways supplements don't replicate. Blueberries don't just dump vitamin C into your bloodstream. They come bundled with signaling molecules that influence gene expression and gut bacteria. The system evolved to digest apples, not capsules.

More importantly, endogenous antioxidant systems—those created by the body itself—are far more powerful. Enzymes like glutathione peroxidase, superoxide dismutase, and catalase are exquisitely tuned. They respond dynamically to need. When a cell experiences oxidative stress, it upregulates these systems automatically. This adaptive response is the key to resilience. But if we artificially suppress ROS with constant supplementation, we deprive the system of its cue to act.

The implications are vast. For decades, public health messaging equated antioxidants with health. But now we know that their benefit depends on timing, dosage, and context. A daily megadose may not just be ineffective—it might be disruptive. The pendulum has swung from panacea to placebo, and in some cases, to poison.

Aging, at its core, isn't just about loss. It's about recalibration. The mitochondria don't need rescue—they need rhythm. Stress, in moderation, becomes a teacher. The free radical, once a villain, now plays a cameo in a more complex story—one about adaptation, not annihilation.

REKINDLING THE FIRE

The gravel crunched under his boots as the mountaineer took another step along the forest path—not the alpine trail of his youth, but a gentle loop behind his daughter's house. At seventyfive, he no longer chased peaks. But his breath was steady, his posture upright, and his steps deliberate. This was his new summit: reclaiming endurance, one quiet walk at a time. He had taken up cycling on an old mountain bike, adjusted his meals to an intermittent fast, and—much to his grandson's amusement—was now popping small blue capsules labeled "NMN."

His sports medicine doctor had smiled at the last checkup. "You're not trying to be thirty," she said, "You're trying to be the strongest seventyfive you can be." For the first time in a decade, he believed that might be possible.

We tend to think of aging as irreversible decline—a one-way slope toward dysfunction. But when it comes to mitochondria, that narrative needs revision. Mitochondria don't just die off with age. They can be rebuilt. They can multiply. They can even get better.

The process is called mitochondrial biogenesis—the birth of new mitochondria from old ones. It's a complex symphony, but its conductor is a molecule called PGC-1α (peroxisome proliferatoractivated receptor gamma coactivator 1-alpha). Discovered in the early 2000s, PGC-1α acts as a master switch, flipping on the genes needed to grow new mitochondria. It's especially active in muscle tissue, heart cells, and brown fat—tissues that demand high energy throughput. When PGC-1α gets going, it not only increases the number of mitochondria but also improves their efficiency. Think of it as replacing flickering bulbs with LEDs while also doubling the number of light fixtures.

And what activates PGC-1α? Not a pharmaceutical giant, but movement. Specifically, aerobic exercise. When muscles contract repetitively—during running, cycling, swimming—

they burn through ATP. The moment the energy supply dips, a cellular alarm goes off. AMPK is triggered. Calcium levels spike. These signals converge to activate PGC-1α, initiating a wave of mitochondrial construction.

Endurance athletes, unsurprisingly, have abundant mitochondria. But what's more encouraging is that older adults can build new mitochondria too. In a landmark study, men and women in their 60s and 70s who undertook regular endurance training saw substantial increases in mitochondrial content in their muscles—comparable, in some cases, to much younger individuals. The key wasn't youth. It was stimulus. Mitochondria are remarkably plastic; they respond to challenge, even in old age.

But movement isn't the only trigger. Fasting, it turns out, also signals cells to ramp up mitochondrial production. When food is scarce, cells interpret it as a stress event and engage survival programs. Among these is mitochondrial biogenesis. From an evolutionary standpoint, this makes sense: when resources are low, the body doesn't shut down—it adapts. It builds a more efficient energy infrastructure to extract every bit of ATP it can from what's available. Just as drought forces roots to grow deeper, fasting forces mitochondria to grow better.

This biological insight has fueled intense interest in NAD+ precursors—compounds like nicotinamide mononucleotide (NMN) and nicotinamide riboside (NR). NAD+ is a coenzyme critical to mitochondrial function; it's like the oil that keeps the engine running smoothly. But as we age, NAD+ levels fall—partly because we use more of it, and partly because certain enzymes consume it more rapidly.

This decline hampers mitochondrial activity, leading to sluggish energy production and increased oxidative stress.

In mouse studies, supplementing with NMN or NR restored NAD+ levels, improved mitochondrial function, and even rejuvenated some age-related declines in endurance, insulin sensitivity, and cognitive performance. While human trials are still ongoing, early results suggest improvements in metabolic health, fatigue, and mitochondrial biomarkers. The mountaineer's NMN pills might be more than hype—they represent the frontier of mitochondrial medicine.

But these are not the only candidates. Molecules like spermidine—a compound found in aged cheese, soy, and wheat germ—have been shown to activate autophagy and mitophagy, potentially clearing out old mitochondria and encouraging new ones. Urolithin A, a gutmicrobiome-derived compound from pomegranate, has shown promise in increasing muscle endurance in older adults, possibly through improved mitophagy. And MitoQ, a mitochondriatargeted antioxidant, is designed to reduce ROS at the source without disrupting beneficial redox signaling.

All these interventions target one truth: mitochondria age, but they can also recover. The decline is not linear. It's responsive.

And then there are the non-chemical tools—elegant in their simplicity, profound in their effects. Heat therapy, such as regular sauna use, has been linked to reduced all-cause mortality in long-term studies. The mechanism? Heat stress activates HSPs (heat shock proteins), which help refold damaged proteins and maintain mitochondrial function. Like fasting and exercise, heat challenges the cell—and the cell responds by toughening up.

Even circadian rhythms matter. Mitochondria have their own internal clocks. Disrupted sleep, irregular eating times, and chronic jet lag can desynchronize mitochondrial function, leading to energy inefficiency and higher ROS output. Conversely, respecting circadian patterns—eating during daylight, sleeping on schedule—keeps mitochondrial cycles running smoothly.

The mountaineer, now months into his revised routine, doesn't chase youth. He embraces adaptation. On good days, he bikes five miles. On harder days, he stretches in the sun. He's found a rhythm: exercise in the morning, food in a short window, early bedtimes. Once a skeptic, he now jokes that he's running a metabolic startup in his cells.

His progress is not measured in VO2 max or personal records. It's in the way his breath no longer trembles when he climbs the stairs. The steadiness of his gait. The clarity in his mind. And the moments—precious and unspectacular—when his body simply works as it should.

Mitochondrial dysfunction was never just about energy. It was about capacity, resilience, momentum. Now, he has all three—not as a gift from youth, but as a prize from effort.

EASY ACCESS ***The whole chapter, in a paragraph***

Mitochondria are the batteries that power every living cell. With age, they leak, accumulate damage in their own DNA, and aren't replaced fast enough by the cell's recycling system. The result is less energy, more inflammation, and the slow shape of frailty. Antioxidant pills don't fix this — they often make it worse, by silencing the very signals that tell the body to adapt. What does fix

> it, at least partly, is the language the cell evolved to listen to: aerobic exercise, fasting windows, regular sleep, intermittent challenge. The mountaineer didn't reverse aging. He rekindled the fire.

He didn't reverse aging. But he did rekindle the fire.

THE UNWELCOME RETIREMENT

CELLULAR SENESCENCE

In a sprawling manufacturing plant on the edge of a mid-century industrial town, there's a man named Walter. Walter used to be a marvel—always on time, I perfectly efficient, the kind of worker who could run three machines at once without breaking a sweat. He trained dozens, maybe hundreds, of new hires, and the floor foremen would whisper his name with reverence when things got tough. But Walter turned 65, then 70. His station changed. He wasn't operating the machines anymore. He wasn't even mentoring. He just... stood around. He didn't clock out. Didn't retire. He loitered near the emergency pull cords, tapping buttons, offering unsolicited advice, and every so often—without reason—he'd set off the fire alarm.

It's not that Walter meant harm. But every time the siren wailed, the whole floor ground to a halt. Nobody was sure what to do with him. He wasn't technically breaking the rules. But he wasn't helping either. And as more Walters showed up—workers who refused to go but no longer contributed—the system began to buckle. Productivity dipped. Tempers flared. The factory, once humming, now wheezed and sputtered.

That factory is your body. Walter is a senescent cell.

Returning to Hayflick and the limit discussed in the opening. In the early 1960s, a young scientist named Leonard Hayflick made a discovery that rattled the foundations of biology. At the time, the consensus among cell biologists was that human cells, once cultured in the lab, could divide forever.

They were thought to be immortal, a notion built on the work of Alexis Carrel decades earlier—a Nobel laureate who had kept chicken heart cells alive for years in a petri dish, claiming that with proper care, cells never had to die.

But Hayflick didn't buy it. His intuition told him something was off. So he ran an experiment with human fibroblasts, the workhorse cells that make connective tissue. He meticulously tracked their divisions, noting every time the population doubled. What he observed was quietly revolutionary: after about 50 doublings, the cells stopped dividing. They didn't die. They didn't turn into other types of cells. They just froze—metabolically active, but permanently arrested. They had reached their replicative limit.

It was a moment of scientific heresy. Hayflick published the results with Paul Moorhead, and the biology community scoffed—at first. Then it repeated his experiments. Again and again, the same result: human cells weren't immortal. They had an internal clock. A counter. And it eventually ran out.

This became known as the Hayflick limit.

What Hayflick had discovered wasn't cell death, and it wasn't temporary hibernation either. It was something stranger: a third state. Not alive in the traditional sense, but not dead. Arrested. Locked in time.

Today, we call that state cellular senescence. A cell enters senescence when it detects serious internal damage—shortened telomeres, broken DNA, oxidative stress, or dangerous oncogene activation. It's the biological equivalent of a self-imposed restraining order: "I've seen enough damage in here. I'm not going to divide anymore. I might become cancerous if I do."

At first glance, that sounds noble. And in many ways, it is. Senescence is a tumor suppressor. It's a safeguard, a cellular brake pedal. When functioning properly, it prevents damaged cells from propagating. In embryonic development, senescence helps sculpt tissues by marking unnecessary cells for removal. During wound healing, it plays a transient role in coordinating repair.

But here's the problem. Senescent cells don't just sit quietly in the corner. They shout.

Once a cell becomes senescent, it starts releasing a cocktail of inflammatory signals, growth factors, and tissue-degrading enzymes. And unlike Walter, senescent cells aren't easy to evict. In youth, the immune system identifies and clears them. But with age, that surveillance breaks down. The cells pile up. Their noise becomes background music, low-grade and chronic. This is what researchers now call inflammaging—a systemic rise in inflammatory signals that wears down organs and accelerates aging.

Senescence is not apoptosis. Apoptosis is clean, precise, and final—a cell recognizes damage and initiates a carefully choreographed suicide. Bits are packed, membranes sealed, immune cells quietly dispose of the body. It's the Japanese tea ceremony of cell death. Senescence, on the other hand, is the loud uncle who won't leave the wedding and keeps heckling the bride. It's not quiet, and it's not dignified. It's a survival strategy that turns sour when left unchecked.

Nor is senescence the same as quiescence—a peaceful pause that cells can exit when conditions improve. Quiescent cells are like workers sent home during a power outage, ready to return. Senescent cells have lost that option. They are locked into idleness, and yet they still occupy space, consume resources, and spread dysfunction.

This understanding reframed how we view aging. It's not simply the result of wearand-tear or slow entropy. Aging, at the cellular level, is often an active process—one the body initiates in an attempt to preserve order, only to be betrayed by unintended consequences. Senescence is a defense mechanism that overstays its welcome.

The question is not whether senescent cells are good or bad. It's when—and how many—and where. A few senescent cells after an injury? Helpful. A spreading crowd of them in your joints, brain, or blood vessels as you hit your seventies? Catastrophic.

Hayflick didn't know it then, but his petri dish fibroblasts weren't just dying quietly. They were setting off alarms. They were foreshadowing a new theory of aging—one that saw time not as decay, but as a negotiation between helpful restraint and harmful stagnation.

And Walter? He's not going anywhere, not unless someone clears him out. But the factory managers have started noticing. Somewhere in a clinical trial room, a new senolytic pill slides across a table to an elderly volunteer. It's the beginning of a new experiment—not just in medicine, but in the management of time itself.

ROTTEN APPLES IN THE BARREL

At first, she thought it was just another pill to ease her knees. The trial paperwork described it as a senolytic cocktail, a compound designed to \"selectively eliminate senescent cells.\" Margaret didn't care much for the jargon. What she cared about was walking up her driveway without feeling like her joints were filled with cement. She had arthritis, the slow kind, not the spectacular kind you get from injury. Just years of wear, of time, of ordinary living.

But what intrigued her most was the metaphor the doctor used: "Think of senescent cells as rotten apples in a barrel. You don't need many to spoil the rest." That image stuck with her.

In the sterile clinic room where she took her first dose, she wondered how many rotten apples she was carrying. And how many had spoiled what was still good.

Cells aren't supposed to live forever. They're supposed to do their job, replicate as needed, and when damaged, either repair themselves or self-destruct. But senescent cells defy that logic. They don't divide, but they don't die either. They linger in a strange limbo—a state scientists call "cellular senescence." These cells emerge after exposure to stress signals: too many divisions, too much DNA damage, or the activation of oncogenes—genes that push a cell toward cancer if left unchecked.

But instead of dying a noble death, senescent cells turn into secretors.

And what they secrete is trouble.

The trouble has a name: the Senescence-Associated Secretory Phenotype, or SASP.

SASP is the toxic language of a cell that's done working but still wants to be heard. Imagine a bitter ex-employee who now stands outside the office every day yelling obscenities about the company to anyone who walks past. That's SASP in biological terms.

The SASP cocktail includes inflammatory cytokines like IL-6 and IL-1β, matrix-degrading enzymes like MMPs (matrix metalloproteinases), and growth factors like VEGF.

These aren't random outbursts—they're highly coordinated molecules that alter the local tissue environment in very specific, very disruptive ways.

First, they erode. MMPs break down the extracellular matrix, the scaffolding that holds tissues together. It's like acid on concrete. In joints, this contributes to the breakdown of cartilage, aggravating conditions like osteoarthritis. In blood vessels, SASP factors weaken vessel walls, accelerating vascular aging and promoting plaques.

Second, they inflame. Cytokines attract immune cells, but not in a way that resolves damage—instead, it creates a low-level, chronic inflammation. The fire never rages, but it never dies either. This state—what researchers call "inflammaging"—has been implicated in nearly every age-related disease, from heart disease to Alzheimer's.

Third, they recruit. SASP can prompt nearby healthy cells to become senescent too—a phenomenon known as paracrine senescence. Like a yawning passenger on a train prompting others to yawn, a senescent cell can induce its neighbors into a similar fate, especially when they're already vulnerable.

That means one rotten apple really can spoil the bunch.

In younger bodies, the immune system patrols for these cells. Natural killer (NK) cells and macrophages act like janitors, sweeping away the dysfunctional ones. But aging blunts that capacity. Immune surveillance weakens, and the janitors start missing shifts.

Over time, senescent cells accumulate like garbage bags never taken out. The numbers are small—usually less than 15% of cells in a given tissue—but their impact is outsized. Because they don't just lie dormant. They shout. And their shouting drowns out the signals of health.

In Margaret's case, it's her knees. But the pattern repeats elsewhere.

In the lungs, senescent cells degrade elasticity, leading to fibrotic diseases.

In fat tissue, they impair metabolic signaling, worsening insulin resistance.

In skin, they reduce collagen production and increase sagging.

In the brain, their inflammatory whispers may prime the nervous system for degeneration.

There's a systemic cost to this cellular chatter.

Margaret's trial used a combination of dasatinib and quercetin—two compounds that had shown promise in mice. In animal models, these drugs selectively cleared senescent cells without harming normal ones. The results were dramatic: improved physical function, better tissue regeneration, even thicker fur. But mice are not humans. And no one knows yet what a long-term senolytic regimen will do in older people.

Still, after four weeks, Margaret noticed something. The pain wasn't gone—but it came later in the day. She could make it through lunch before the ache started. She was no fool. It could be placebo. Or it could be her body, quieter now that some of the rotten apples had been tossed.

What makes senescence uniquely cruel is its subtlety. The cells don't multiply or form tumors. They don't bleed or burst. They don't show up on scans. But they speak, chemically, to the cells around them. And what they say encourages decay.

There's something almost poetic about it. Cells that once guarded the body—by refusing to divide when damaged—end up degrading it. They were the heroes of youth, now turned saboteurs in age.

And yet, understanding that shift opens the door to new interventions.

Scientists now use SASP profiles as biomarkers, trying to map which tissues are most poisoned and when. Clinical trials are underway in osteoarthritis, pulmonary fibrosis, and even cognitive decline. The hope is simple: remove the source of inflammatory noise, and let the body regenerate without distraction.

But the challenge remains—identifying which cells are senescent and ensuring senolytics don't harm the ones playing essential roles.

Because not all senescent cells are harmful all the time. In wound healing, they actually help. In embryonic development, they sculpt tissues. The line between beneficial and harmful is context-dependent, blurry, and dynamic.

For now, Margaret walks the line—one foot in the world of possibility, the other still grounded in pain.

She doesn't know if the drug is working.

But she knows this: the shouting in her joints has quieted. A little.

And maybe, just maybe, one fewer rotten apple is spoiling her barrel.

The Good, the Bad, and the Necessary

Before senescence was cast as the villain of aging, it was the sculptor of form and a quiet emergency brake on cancer. The paradox of cellular senescence is that it was born as a protector. Only later did it grow unruly.

In the 2010s, a group of Spanish scientists engineered a mouse embryo that couldn't trigger senescence. They weren't testing aging—they were studying development. What they saw was stunning: malformed limbs and tangled tissue structures. The embryos had too many cells, but fewer distinctions. Fingers didn't separate properly. Eyes were misshapen. Without senescence, the blueprint went haywire. It was as if the construction crew forgot when to stop.

We tend to think of aging as the breakdown of mechanisms. But senescence was never just breakdown. It was built into the earliest scripts of life.

Embryonic development may seem like a flurry of growth—cells dividing, differentiating, forming organs and limbs—but it also involves removal, restraint, and revision. Senescent cells emerge in specific locations, help remodel tissues, and then vanish.

In the embryonic limb, for instance, senescence prunes excess cells between forming digits. Fingers are not simply built; they're carved. In the kidney, senescent cells help shape tubules by secreting growth-regulating factors. In the inner ear, in the eye, in the brain—senescence orchestrates the withdrawal of certain cells so that other cells may organize more precisely.

What makes it more elegant still is that these senescent cells are programmed to be cleared. They arise, contribute, and then disappear, much like construction scaffolding taken down once the building is complete.

This phenomenon is called developmental senescence, and it follows a strict choreography. The temporary senescent cells do not express inflammatory SASP components, and their clearance is managed swiftly by embryonic immune cells. The problem with aging is not that senescence exists—but that the cleanup crew retires too soon.

Move from embryo to injury, and the story repeats with a twist. A skin wound activates local fibroblasts. Some of them become senescent. These cells start pumping out factors that stimulate nearby stem cells and orchestrate tissue regeneration. They're like field medics coordinating emergency response.

Senescent cells in wounds recruit immune cells to the site, secrete matrix-modifying enzymes to clear debris, and secrete platelet-derived growth factor AA (PDGF-AA), which promotes the regrowth of skin. Once the job is done, the senescent cells are removed—either through immune clearance or by self-destructing.

Studies show that if you prevent senescent cells from forming during wound healing, the tissue heals more slowly and less effectively. If you artificially prolong their presence, the wound site becomes fibrotic and inflamed. It's a delicate balance: senescent cells are a signal, not a solution. Their value lies in their brevity.

Senescence also acts as a guard against cancer. When a cell detects oncogenic stress—a mutated RAS gene, for instance—it often responds by triggering senescence. This is a failsafe mechanism. Instead of dividing uncontrollably and forming a tumor, the cell halts itself permanently. Its division program is locked.

This phenomenon is well-documented in precancerous lesions. In earlystage skin, lung, and liver growths, senescent cells can often be found forming a boundary—like sentinels barring further progression. It's part of what's called oncogene-induced senescence (OIS), and it reflects the deep evolutionary tension between growth and control. We need cells to multiply during youth and healing, but we also need brakes. Senescence is one of the oldest brakes in our genetic system.

In mice genetically engineered to bypass senescence, tumor formation accelerates. In contrast, boosting senescence in tumorprone mice can delay or suppress cancer. In this light, senescence is not a malfunction—it's a failsafe against worse outcomes.

Senescent cells have also been found to limit tissue scarring in some contexts. Take the liver. After an acute injury, senescent hepatic stellate cells secrete enzymes that degrade excess collagen and prevent runaway fibrosis. They also produce anti-fibrotic molecules like matrix metalloproteinases (MMPs), which dissolve early scar tissue.

This behavior has been observed in models of lung and kidney fibrosis too. When senescent cells are absent or dysfunctional, fibrosis tends to worsen. The current belief is that senescence provides a kind of stopsignal, halting cells that would otherwise over-repair the tissue, laying down excess matrix that stiffens and impairs organ function.

Here again, context is everything. Senescence is protective when tightly timed, damaging when prolonged. The trouble in aging is not that senescent cells exist—but that they're not cleared. They persist, spewing signals long after their task is done.

This is the crux of the paradox. Senescence evolved for good reasons: to shape the body, heal wounds, and prevent cancer. But like a safety net left up too long, it begins to catch the wrong things.

As we age, the immune system loses its sharpness. The natural killer cells that used to prune senescent cells become sluggish. The surveillance declines. Senescent cells accumulate in joints, fat, muscle, liver, even brain. And the very features that once made them useful—their ability to signal, to stop growth, to alter their environment—become liabilities. This sets up a key caution for therapies aimed at clearing senescent cells. If you remove all senescent cells, do you impair wound healing? Do you increase risk of fibrosis or even certain tumors? The truth is, we don't fully know yet. That's why senolytic drug research is advancing slowly and deliberately.

It's tempting to label senescent cells as either good or bad, but the real challenge is managing their timing and context. Like fire, they warm and destroy depending on the boundaries.

Senescence is not a flaw—it's a feature. The problem is duration. What should be a temporary adaptation becomes a chronic burden. If we understand senescence not as a disease but as a tool gone awry, we get closer to intelligent interventions.

In this chapter on aging, senescence represents the blurred line between biology's brilliance and its unintended consequences. We were sculpted by senescence. We were saved by it. And now, it might be killing us—slowly, by refusing to leave the stage.

Next comes the more controversial question: if we can remove these cells... should we?

THE HUNT FOR ZOMBIE KILLERS

The woman had signed up out of curiosity more than hope. Her osteoarthritis had stiffened her knees for years—mornings were the worst, when the joints moved like rusted hinges. Her grandson helped her into her walking shoes now. She missed her independence.

The trial pamphlet had used unfamiliar language: "senescent cell clearance," "inflammatory mediators," "potential functional improvement." But what caught her attention was the nickname whispered by one of the younger staff members: zombie cells. She smiled at the absurdity. A pill to hunt the undead in her knees? Still, what did she have to lose?

Three weeks in, she noticed it first when getting out of bed. The sharp morning pain was still there, but duller, somehow. Not gone. But less defining. By week five, she was walking to the post office again—slowly, but unassisted. The researchers called it "subjective improvement." She called it getting her legs back.

Across the globe, scientists were hunting zombies in petri dishes and mouse cages.

It started, fittingly, with mice who wouldn't grow old the way they were supposed to.

In one groundbreaking experiment, researchers engineered mice to carry a genetic kill switch. The switch could be activated to eliminate cells expressing p16\^INK4a, a classic marker of senescence. As these mice aged, scientists flipped the switch—and watched. The results were electric. Compared to their untreated siblings, these mice had healthier kidneys, stronger grip strength, thicker fur, and even delayed cataracts. Their median lifespan increased by up to 30%.

It was one of the most compelling demonstrations that senescent cells were not just passengers in aging—they were drivers.

The logic was elegant. If senescent cells accumulate with age and disrupt tissue function through the SASP, then perhaps aging itself could be slowed by simply taking them out. But elegant ideas are rarely simple in biology. The leap from genetically engineered mice to pill-based therapy in humans required a new class of drugs: senolytics.

Senolytics are compounds that selectively induce death in senescent cells while sparing healthy ones. The term entered the scientific mainstream in the mid-2010s, when researchers at the Mayo Clinic screened various molecules and hit on a powerful combination: dasatinib, a cancer drug, and quercetin, a plant flavonoid found in apples and onions. Together, they acted like a targeted cleanup crew. Senescent cells, due to their altered survival pathways, were uniquely vulnerable to this cocktail.

In cell cultures and mice, the results were striking. After treatment, not only were senescent cells reduced, but inflammation markers dropped, and tissue function improved. It wasn't a miracle cure—aging wasn't reversed—but it was as if the body's environment became friendlier, less toxic. A chronic burden had been lifted.

In early human studies, small and carefully monitored, similar patterns began to emerge.

In one trial involving patients with idiopathic pulmonary fibrosis—a brutal, progressive lung disease—participants took a course of dasatinib and quercetin for three weeks. The primary goal was safety, but some patients showed modest improvements in physical performance. No serious side effects emerged, sparking interest in further trials.

In another small pilot study, older adults with type 2 diabetes and kidney dysfunction received the same cocktail. Again, the treatment appeared to reduce markers of senescence in fat and skin biopsies. The trials were far from definitive, but they whispered a tantalizing possibility: that some aspects of aging might be therapeutically malleable.

For osteoarthritis, where senescent cells accumulate in cartilage and synovium, researchers hoped that reducing SASP-driven inflammation might ease joint pain and slow disease progression. That was the trial the woman had joined. She wasn't told if she had the drug or placebo—it was double-blinded. But her knees had a say in the matter, and they were speaking with more ease each week.

Yet the road forward is not without obstacles.

One challenge is targeting. Senescent cells are heterogeneous—there's no universal marker. They can arise from different triggers and express different surface proteins, depending on their origin and tissue type. Dasatinib and quercetin are blunt tools, targeting only certain kinds of senescent cells effectively. Other senolytics, such as fisetin (another plant-derived compound) and navitoclax (a BCL-2 inhibitor), have shown promise, but also come with limitations. Navitoclax, for instance, can cause platelet toxicity—a nonstarter for many elderly patients.

Another issue is timing and duration. How often should senolytics be administered? Monthly? Annually? Should they be preventive or reactive? Unlike antibiotics that clear an infection, the optimal "dose and duration" of cellular rejuvenation is still murky. Overzealous removal of senescent cells might, paradoxically, impair processes like wound healing or immune modulation. Context matters. The therapy must be dialed in, not blasted across the body indiscriminately.

Then there's the delivery puzzle. Reaching senescent cells in deep tissues like the brain or joints requires molecules that can cross biological barriers without being rapidly degraded or accumulating offtarget effects. Nanoparticle-based delivery, antibody-guided targeting, and prodrug approaches are all being explored. But they are in early stages.

Finally, there is the haunting specter of long-term consequences. What happens when we alter the body's natural surveillance systems? Senescence evolved for a reason—to stop runaway cell division, to sculpt embryonic tissues, to limit fibrosis. Will suppressing it leave other vulnerabilities exposed? Could cancer risks rise if senescence fails to check pre-malignant cells? Might senescent cells in some contexts be necessary?

These aren't trivial concerns. They are reminders that biology is never simple. Aging is not a bug to be fixed; it's a program to be understood.

Still, the woman in the trial is walking more. She smiles more. Her doctor says the data are still coming in, but she feels the truth in her knees. It might not be magic, but it feels like motion.

And motion, at her age, is everything. Mixed Signals The headline was too neat to be true. "Zombie Cells Cleared, Life Extended!" splashed across a major science outlet.

The first studies were nothing short of miraculous—at least on the surface. Mice engineered to clear senescent cells lived longer, looked younger, and performed better on physical tests. Their fur grew thicker. Their kidneys filtered more efficiently. Their hearts pumped with renewed vigor. The public heard \"anti-aging.\" But in the data, there was a quieter distinction that told a more complicated tale: these mice lived longer in the middle of life—not necessarily at the end.

That is, their median lifespan—the age by which 50% of a group has died—was significantly improved. They stayed healthier longer. But the maximum lifespan, the age reached by the longestliving mouse, barely budged. This pattern raised a critical question: were senolytics improving healthspan without fundamentally altering the upper boundaries of aging?

It turns out, biology doesn't offer upgrades without caveats. One of the most counterintuitive findings came when researchers tried giving senolytics to very old mice—already past their prime. In some cases, instead of rejuvenation, the intervention triggered declines. Frail animals became more vulnerable, and in some tissues, degeneration seemed to accelerate. The paradox? Clearing senescent cells late in life may tip already-precarious systems into chaos. The timing, it appeared, mattered as much as the treatment. There was also growing concern about unintended side effects. In youthful settings, senescence plays roles in healing wounds and managing tissue repair. When scientists began clearing senescent cells indiscriminately, some animals healed more slowly. Skin lesions lingered. Tissue regeneration stuttered. It wasn't hard to see why: senescent cells, though noisy and inflammatory, often released molecules that coaxed immune cells or growth factors into action. Their total removal was like muting the fire alarm without checking if the building was still burning.

Then there's the issue of cellular diversity. Senescence is not a monolith. Some senescent cells are highly inflammatory, spewing SASP like a broken hydrant. Others are quiet, possibly even beneficial. Senescent cells in the liver differ wildly from those in fat tissue or cartilage. Even within a single organ, their molecular signatures—what genes they express, what molecules they release—can vary widely.

This heterogeneity makes blanket senolytic therapy a risky bet. A drug designed to target one type of senescent cell might miss others—or worse, affect healthy cells. Imagine trying to evict troublemakers from a crowd based only on a single trait—say, wearing a red shirt. You might catch a few vandals, but also a lot of bystanders.

Detecting these cells isn't simple either. Scientists often use markers like p16\^INK4a or β-galactosidase to flag senescence. But no universal marker exists. Some senescent cells don't express these hallmarks at all. Others do, but for unrelated reasons. Without a clean fingerprint, therapies operate partly in the dark. It's like trying to design a sniper rifle without knowing what your target looks like.

Even at the population level, there are signs of complexity. Some senescent cells, oddly enough, may protect against fibrosis by halting the runaway growth of fibroblasts—cells that create scar tissue. If senescence reins in this process, then removing those cells could worsen scarring, not improve it. In one study, the clearance of senescent fibroblasts in the liver led to increased fibrosis, not less. The balance is delicate.

Back in the clinical trial, our participant is feeling better. Her knees ache less. She walks to the market without stopping. But when the researchers scan her joints, they don't see a dramatic reversal—just a slowing, a softening of the inflammation. It's not a miracle. It's a dial turned slightly back, not a switch flipped off. And that's the key insight.

Senescence is not a simple villain to vanquish. It's a program—ancient, conserved, and deeply woven into the biology of life. In youth, it protects. In middle age, it manages damage. In old age, it accumulates, turning from helper to hazard. And like many things in aging, it falls under the principle of antagonistic pleiotropy—a trait that helps early in life but harms later. The same fire alarm that keeps the building safe can become a nuisance when no one's checking the flames.

So what's the future? Not scorched-earth senolytics. But precision clearance—targeting only the most harmful cells, at the right time, in the right tissues. Think of it as cellular diplomacy, not war. Tools like senomorphics—drugs that mute SASP without killing the cell—may be gentler alternatives. Others are exploring transient senolysis: periodic pulses rather than continuous treatment, giving the body time to recover and recalibrate.

In aging, as in life, more isn't always better. Sometimes, the answer isn't to erase, but to edit. To nudge rather than nuke. As one researcher said after reviewing the latest data: "Senescence is a conversation, not a monologue. We just need to learn when to listen—and when to quiet the noise."

Antagonistic Pleiotropy and the Price of Protection

At first, the idea seems cruel. That what saves us early in life might betray us later. That the very defenses evolution built into our bodies—like a well-placed firewall—can smolder into something more like arson. But biology has always been a game of tradeoffs, and in the realm of aging, none is more poignant than the concept of antagonistic pleiotropy.

It's a mouthful. But its meaning is beautifully simple—and devastating. In 1957, evolutionary biologist George C. Williams proposed this nowseminal idea: that certain genes or traits offer benefits in early life, especially in terms of reproduction or survival, but carry costs later in life. And because evolution doesn't particularly care what happens after you've reproduced, those latelife costs are tolerated. Hidden. Deferred. Like debt accumulated by youth, and paid back with interest by age.

Senescence, as it turns out, may be one of those debts.

Take a single cell exposed to an oncogene—a mutation that would normally send it careening toward cancer. The body responds brilliantly. The cell slams on the brakes. Enters senescence. It's not dead, but it's done dividing. It can no longer proliferate into a tumor. This is good. Life-saving, even. It's why senescence is often hailed as a tumor-suppressive mechanism. In young bodies, it helps keep us safe.

But fast-forward a few decades. That cell, and thousands like it, have lingered. They haven't been cleared out by the immune system, which now itself has grown sluggish. They begin to secrete inflammatory signals—cytokines, growth factors, enzymes—that degrade the surrounding tissue. They don't work. They won't leave. And they don't stay quiet.

This, in essence, is the price of protection. The bargain of antagonistic pleiotropy: protection from cancer at twenty, chronic inflammation at seventy.

In the clinical trial we've followed throughout this chapter, the elderly woman who entered with aching knees continues to show signs of progress. She walks with more ease now. Her pain has softened into something bearable. But the researchers are not popping champagne. They're watching the other shoe. Because the clearance of senescent cells—especially when done pharmacologically—might come with side effects that don't show up immediately. Could it impair wound healing? Could it remove senescent cells that, in certain contexts, were still serving useful functions? These aren't hypothetical worries. They're part of the emerging, complicated terrain of senescence science.

You see, evolution doesn't design for comfort. It designs for reproductive fitness. Once you've passed on your genes, evolution more or less steps aside. And senescence, with all its doubleedged consequences, fits this design perfectly. It protects the young at the cost of the old. It buys short-term survival by mortgaging long-term vitality.

This theme—of biology offering gifts laced with hidden expiration dates—repeats across aging. The immune system is ferocious in youth, but overactive in old age, tipping into autoimmunity. Bone remodeling is dynamic in youth, but later, the same mechanisms contribute to osteoporosis. Even metabolism—so efficient in early life—can give way to insulin resistance and metabolic syndrome. It's as if every protective wall becomes, over time, a confining box.

Some scientists wonder if this evolutionary tradeoff was inevitable. After all, in a natural world without antibiotics, hip replacements, or retirement homes, few individuals lived long enough for these latent costs to matter. The species advanced. The price of survival was paid by individuals, quietly and invisibly, through senescence and its ilk.

But now, with lifespans stretching beyond what evolution ever accounted for, those costs are coming due. And the question is no longer whether senescence is a problem—but whether we dare rewrite the bargain.

There are ethical dimensions here, too. If senolytic drugs become viable, who gets access? If clearing senescent cells can extend healthspan—or even lifespan—how does society adapt to an even older population? What happens when aging, long considered inevitable, becomes modifiable for some and not for others?

Senescence, after all, isn't just a glitch. It's a feature. A feature with unintended consequences. A gift wrapped in barbed wire. The chapter closes here—not with triumph, but with motion. If we can clear the saboteurs of aging... should we?

Reader-tool boxes for this chapter

CLOSER LOOK ***What a senescent cell actually does***

Cellular senescence is a stable cell-cycle arrest triggered by various insults — telomere attrition, DNA damage, oncogene activation, oxidative stress. The cell stops dividing, but it does not die. Instead it adopts a distinctive secretory profile: the senescence-associated secretory phenotype, or SASP. Senescent cells pump out a cocktail of inflammatory cytokines (IL-6, IL-8), proteases, and growth factors that affect every neighbor within reach.

In a young tissue, the SASP serves a purpose. It recruits immune cells to clear the senescent cell out, helps remodel tissue during wound healing, and limits the spread of damaged or precancerous cells. The trouble starts when the immune system itself ages. Senescent cells accumulate, their SASPs persist, and what was a transient repair signal becomes chronic background inflammation — driving fibrosis, vascular stiffness, joint degeneration, and metabolic dysfunction in tissues nowhere near the original injury.

Coppé et al. (2008). Senescence-associated secretory phenotypes reveal cell-nonautonomous functions of oncogenic RAS and the p53 tumor suppressor. PLoS Biology, 6(12), e301.

CLOSER LOOK ***INK-ATTAC — the experiment that broke the field open***

In 2011, Darren Baker and Jan van Deursen at the Mayo Clinic engineered a transgenic mouse called INK-ATTAC. The animal carried a suicide gene under control of the p16Ink4a promoter — meaning any cell that turned on p16 (a senescence marker) could be killed on demand by giving the mouse a small molecule. They tested the system in a progeroid mouse model that ages rapidly.

Selectively eliminating p16-positive cells delayed the onset of cataracts, sarcopenia, and fat tissue loss. Even better: late-life clearance, started after the mice were already showing age-related decline, slowed the progression of the deficits. For the first time, the field had a causal demonstration that senescent cells were not just markers of aging but drivers of it. The next decade of senolytic drug development started in that paper.

Baker et al. (2011). Clearance of p16Ink4a-positive senescent cells delays ageing-associated disorders. Nature, 479(7372), 232–236.

CLOSER LOOK ***Dasatinib + quercetin — the first translational senolytic***

The first senolytic combination to make it into human trials was dasatinib (a tyrosine kinase inhibitor used in leukemia) plus quercetin (a plant flavonoid). The pairing was identified in 2015 by James Kirkland's group at Mayo, who reasoned that senescent cells survive their own SASP-induced toxicity by activating specific anti-apoptotic pathways. Drugs that block those pathways should preferentially kill senescent cells.

In 2018, Xu and colleagues showed that dasatinib + quercetin given to old mice improved physical function and increased remaining lifespan. The first human pilot trial, published in 2019, treated patients with diabetic kidney disease using a brief 'hit-and-run' protocol — three days of treatment, no dosing for the rest of the month — and demonstrated reductions in senescent cell burden in adipose tissue and skin. The trials are early, the patient numbers small, the long-term safety unknown. But for the first time, a drug protocol designed around aging biology has cleared the bench-to-bedside gap.

Hickson et al. (2019). Senolytics decrease senescent cells in humans: Preliminary report from a clinical trial of Dasatinib plus Quercetin in individuals with diabetic kidney disease. EBioMedicine, 47, 446–456.

EASY ACCESS ***Why we have senescence at all***

Senescence didn't evolve to age you. In a cut on your skin, senescent cells appear within hours, secrete growth factors that recruit fibroblasts, help close the wound, and are cleared out by the immune system within days. The same machinery that becomes destructive in old tissue is genuinely useful in young tissue. The problem isn't senescence itself; it's the failure to clean up afterwards. As immune surveillance wanes with age, transient retirements become permanent occupations, and useful repair signals become chronic inflammation. Senolytics try to do, pharmacologically, what your young immune system used to do biologically.

KEY TERMS ***Vocabulary of the senescence chapter***

Senescence — Stable cell-cycle arrest in response to stress. The cell stops dividing but does not die.

SASP — Senescence-associated secretory phenotype. The inflammatory cocktail secreted by senescent cells.

p16Ink4a — A cyclin-dependent kinase inhibitor. The most reliable molecular marker of senescent cells in tissue.

Senolytic — A drug that selectively kills senescent cells. Dasatinib + quercetin and fisetin are the prototypes.

Senomorphic — A drug that suppresses the SASP without killing the senescent cell. An alternative therapeutic strategy.

RENEWABLE PARTS RUNNING LOW

STEM CELL EXHAUSTION

In the town of Guildford, tucked behind a sagging willow tree and a half-collapsed wooden sign, once stood a thriving carpenter's workshop. The place had rhythm. Sawdust floated through sunbeams like golden snow. Wood shavings curled like ribbon onto the floor. The master craftsman—a man with calloused hands and a soft, impatient heart—rarely worked alone. He had six apprentices, each with a different strength: one could whittle to perfect symmetry; another could match grains from memory. They didn't just repair broken chairs or creaky floorboards. They revived wood as if it had memory, carving life into age.

But over the years, the shop grew quieter. One apprentice left for the city. Another fell ill. Eventually, the master came in alone. His hands still moved with skill, but not speed. Nails bent. Joints misaligned. Where once there was regeneration, now there was delay—and, sometimes, failure. The workshop hadn't collapsed. But it no longer healed as it used to.

This is not just a tale about wood and time. It is the biology of aging, rendered in metaphor. The carpenter is your body. The workshop is your tissue. And the vanished apprentices? They are your stem cells—once vibrant, now exhausted or gone.

Every tissue in the human body has its own version of these cellular apprentices. Hematopoietic stem cells in your bone marrow constantly replenish the blood—producing red cells that carry oxygen, white cells that fight infection, and platelets that prevent you from bleeding out from a paper cut. Muscle satellite cells sit dormant beside muscle fibers, ready to spring into action when you tear a hamstring or lift something too heavy. Neural progenitor cells quietly maintain certain parts of the brain's architecture, nudging new neurons into being under the right conditions. And mesenchymal stem cells are the unsung sculptors of bone, cartilage, and connective tissue—materials that make up the scaffolding of your body.

These cells are not immortal. They're not infinite. But they are, in youth, astonishing. They divide. They respond to signals. They repair. They are the reason a six-year-old's scraped knee scabs over in a day, while an eighty-year-old's bruised leg might darken, swell, and linger.

What changes?

Two things. First, the obvious: numbers. There are simply fewer stem cells with age. Bone marrow stem cells dwindle. Muscle satellite cells decline. The thymus—a school for training immune cells—shrinks, starving the body of fresh T-cells. This attrition alone reduces the regenerative pool. But the more subtle shift is qualitative. Even when stem cells remain, they often lose their responsiveness. They don't divide when they should. Or they divide badly. Their ability to differentiate into specialized cells fades. They've not died—but they've grown quiet, stubborn, unpredictable.

This is stem cell exhaustion. Not death, but dormancy.

Not absence, but apathy.

The scientific definition captures this nuance. A stem cell is an undifferentiated cell capable of self-renewal and of generating one or more specialized cell types. In practical terms, they're the multipurpose repair kits for each tissue. But aging impairs both aspects: their self-renewal slows, and their capacity to specialize degrades.

What triggers this deterioration? Multiple assaults, accumulated over decades. DNA damage—from radiation, oxidative stress, or replication errors—can push stem cells into senescence, a kind of forced retirement. Telomere shortening, which acts like a cellular clock, tells some stem cells it's time to stop dividing. Chronic inflammation, increasingly common with age, alters the "niche"—the microenvironment around stem cells—making it hostile or confusing. Even metabolic changes can skew stem cell behavior, tilting hematopoietic stem cells, for instance, toward producing more myeloid cells (like macrophages) and fewer lymphocytes (like B and T cells), subtly reshaping the immune landscape.

But the exhaustion is not uniform. Nor is it fully irreversible.

Some tissues age faster than others. Muscle stem cells seem more resistant in certain conditions than neural ones. And even within a tissue, some stem cells may retain youthful function if shielded from damage or placed in a supportive environment. This heterogeneity is critical—and hopeful. It suggests stem cell exhaustion isn't just an unavoidable ticking clock. It's a response. And responses can, sometimes, be reversed.

One of the earliest clues came from transplantation studies. Scientists took hematopoietic stem cells from young mice and injected them into old mice. Remarkably, these old recipients began producing more balanced, youthful blood cells. The inverse experiment—placing old stem cells into young mice—often led to sluggish or skewed blood production. The implication was clear: stem cells age, and aged stem cells carry that aging with them then came the twist.

In a series of experiments that would later be immortalized in scientific lore, researchers stitched the circulatory systems of old and young mice together—a procedure known as heterochronic parabiosis. The old mice, now sharing blood with the young, began to change. Their muscles healed faster. Their livers regenerated better. Even their brains showed increased neurogenesis. It was as if the presence of youthful blood had whispered to their stem cells: Wake up.

These findings didn't just suggest that aging was reversible at a functional level. They hinted that many stem cells might not be lost at all. They were dormant. Inhibited. Suppressed by environmental cues—perhaps inflammatory molecules, metabolic signals, or epigenetic silencing. But not broken.

Which brings us back to the carpenter.

He didn't forget how to build. His tools still work. But without apprentices—without that renewing workforce—his shop can't repair at scale. He can patch a stool, not rebuild a staircase. And yet, what if the apprentices hadn't truly vanished? What if they stood outside the door, confused by a sign that read "Closed" when it should've said "Help Needed"? What if the master, given the right signals, could call them back in?

That's the central mystery stem cell biology now faces. Are we witnessing true depletion? Or a widespread miscommunication between our bodies and their builders? The distinction matters—immensely. Because if stem cell exhaustion is reversible, not terminal, then aging itself becomes a conversation we can re-enter. A dialogue we might one day edit.

But for now, the workshop stands halfstaffed. The repairs come slower. The cracks widen. The master carries on. He's not out of time. But time, for the first time, feels finite.

FRACTURES THAT WON'T HEAL

We take healing for granted—until it slows to a crawl. Bone, like all tissues, depends on a dynamic cycle of damage and renewal. Fractures trigger a cascade: blood vessels break, inflammation follows, immune cells swarm, and finally, stem cells are recruited to rebuild what's broken. In youth, this is an orchestra. In age, it's a scattered soloist playing out of tune.

At the heart of this process are mesenchymal stem cells (MSCs)—multipotent cells that live in the bone marrow and can turn into osteoblasts (bonebuilding cells), chondrocytes (cartilage), or even adipocytes (fat). But aging distorts their priorities. MSCs begin to favor adipocyte differentiation—resulting in more marrow fat and less bone. With age, their number drops, and their responsiveness to injury signals fades.

This is not just an internal problem. The stem cell niche—the microenvironment that nurtures and guides stem cells—also degrades. Blood vessels become stiff and leaky, producing fewer rejuvenating signals like VEGF. Systemic inflammation—termed "inflammaging"—sends confusing messages, often blocking the normal repair sequence. MSCs become "conflicted"—less inclined to mobilize, less able to rebuild.

Wound repair in aged muscle follows a similar script. Muscle satellite cells, the tissue's resident stem cells, remain dormant in a healthy adult until injury occurs. Then they activate, divide, and rebuild fibers. But with aging, these cells are fewer, slower, and often enter senescence. Even when they're still there, they may fail to respond to repair signals. A scratch becomes a scar. A sprain becomes stiffness for life.

The immune system plays a supporting—but essential—role. Macrophages help coordinate the transition from inflammation to regeneration. In older individuals, this balance is off. The wrong macrophage subtypes dominate, prolonging inflammation and suppressing repair. It's not just a stem cell problem; it's an ecosystem issue.

We see the consequences every day. Older adults with fractures have a higher risk of complications, longer hospital stays, and lower functional recovery. A hip fracture, in particular, is a catastrophic event—not because of the break, but because of what it reveals: the engine of repair has run low on fuel. No matter how skilled the surgeon, there's only so much you can do if the body's workers don't show up.

AGED BY CHEMOTHERAPY

Meet Wally. Sixtytwo, recently retired, recently diagnosed. Non-Hodgkin lymphoma. His oncologist maps out the treatment plan. The chemotherapy—strong, aggressive, lifesaving.

And it works. The tumor shrinks. But months later, Wally is anemic, bruising easily, and constantly fatigued. His cancer is in remission. But something else is broken. His blood.

This, of course, is not an uncommon story. Chemotherapy, for all its benefits, is bluntforce trauma at the cellular level. It targets fastdividing cells, and unfortunately, the bone marrow—home of hematopoietic stem cells (HSCs)—is collateral damage.

HSCs are the progenitors of every blood and immune cell in our body. Red cells, platelets, neutrophils, B and T lymphocytes—they all trace their lineage to this small, precious population. In youth, HSCs replenish the blood system daily with remarkable fidelity. But with age, they become fewer, and more dysfunctional. They increasingly favor the myeloid lineage (inflammatory white cells like monocytes and granulocytes) over the lymphoid lineage (adaptive immunity). The result? Weaker defense against infections. Poor vaccine responses. Rising inflammation.

Aging also brings another twist: clonal hematopoiesis of indeterminate potential (CHIP). As HSCs dwindle, a few mutant clones begin to dominate blood production. These clones often carry mutations in genes like DNMT3A or TET2—linked to leukemia. While most people with CHIP don't develop cancer, the condition does raise the risk. It also increases cardiovascular risk, likely via systemic inflammation.

Chemotherapy acts like an accelerant. It damages the bone marrow further, reducing the regenerative capacity and making clonal dominance more likely. In some cases, the marrow never recovers—patients develop therapy-related myelodysplastic syndromes or aplastic anemia, conditions requiring bone marrow transplants.

Wally's doctors give him a stem cell transplant—either autologous (his own stem cells harvested before chemo) or allogeneic (from a donor). This procedure, while transformative, is also revealing. It shows how utterly reliant our life systems are on stem cells. Without HSCs, we have no blood. No immunity. No oxygen transport. No clotting. We aren't just weakened—we cease to function.

Hematopoietic stem cell exhaustion is perhaps the most dramatic demonstration of aging's cost. It's not hypothetical.

It's observable in lab counts, immune responses, and recovery times. And it isn't just about the number of cells—but their resilience. Older HSCs are more sensitive to stress. They harbor more DNA damage. Their telomeres are shorter. Their repair enzymes are slower. In short, they age just as we do—but their decline is catastrophic because they carry the future of all other cells.

DORMANT, NOT DEAD

In 2005, a strange and slightly macabre experiment stunned the aging research community. Scientists at Stanford stitched together two mice—one young, one old—so that they shared a common bloodstream. The technique, known as heterochronic parabiosis, wasn't new. But what the old mouse did in the days and weeks after joining circulatory forces with its youthful partner was breathtaking. Its muscles regenerated faster. Its liver showed signs of repair. Even neural stem cells in its brain, long considered past their prime, stirred as if roused from hibernation. None of the old mouse's DNA had changed. What had changed was its environment. And that single fact reframed one of the most basic assumptions of aging: maybe some of our decline isn't due to irreparable damage—but to silence.

This reframing carries special weight when we look at stem cells in aged muscle, a system that has long stood at the heart of regeneration science. In youth, muscle satellite cells—the body's ondemand repair crew—sit just beneath the muscle membrane, inert until injury or stress signals awaken them. They proliferate, differentiate, and rebuild damaged tissue with military-like precision. A pulled hamstring, a scraped knee—these were training exercises. But as age advances, these same satellite cells become quieter. Not absent, not obliterated. Just\... unresponsive.

At first glance, scientists believed these cells were simply dying off. That fit the broader narrative of depletion. But closer inspection—especially using lineage tracing and transplantation experiments—painted a subtler picture. Many satellite cells remain in old muscle. Their numbers drop slightly, yes, but the bigger change is qualitative. They're present. They just don't act.

The reason lies not in the stem cells themselves, but in the shifting terrain that surrounds them. This terrain—the \"niche,\" as biologists call it—is the microenvironment that provides stem cells with the signals they need to survive, divide, and differentiate. Like a radio that needs the right frequency to receive a message, stem cells depend on specific molecular cues. And in aged tissue, the frequencies shift.

Inflammation is one culprit. Chronic low-grade inflammation—what scientists call "inflammaging"—bathes the stem cell niche in a fog of cytokines and immune noise. Signals like IL-6 and TNF-alpha, beneficial in short bursts, become inhibitors over the long term, distorting the stem cell's internal compass. Instead of sensing a need to act, the cells detect a confusing and often contradictory message: danger everywhere, but no clear mission. So they stay put.

Another layer of inhibition comes from altered cell

signaling pathways. In particular, the Notch and Wnt signaling axes—two pathways critical for activating and regulating stem cells—undergo dramatic shifts with age. Notch signaling, essential for keeping muscle stem cells in a quiescent but ready state, declines. Wnt signaling, which encourages differentiation, becomes hyperactive in some contexts, prematurely pushing stem cells into differentiation before they can proliferate. The result is a dysfunctional repair loop: stem cells either sit dormant or exit the cycle too soon.

There's also the problem of diminished growth factor availability. Molecules like fibroblast growth factor (FGF) and insulin-like growth factor (IGF-1), which normally prime stem cells for activation and tissue repair, become less abundant with age. Even when present, aging tissues often express fewer receptors to respond to them. It's as if the workshop is receiving fewer orders—and has also lost the ability to read the ones that do arrive.

On top of that sits the increasingly chaotic layer of epigenetic drift. Over time, the patterns of chemical tags that regulate gene expression become noisier. These changes don't alter the DNA sequence, but they do shift which genes are turned on or off. In stem cells, this can lead to the suppression of regenerative programs and the activation of stress responses. It's not that the instructions are wrong. It's that the instruction manual has become disorganized and harder to read. Aged stem cells show more DNA methylation at youthful gene loci and less acetylation at those needed for regeneration, locking them into a silent state.

The best evidence that these cells are not lost but latent comes from the heterochronic experiments. In one study, muscle stem cells isolated from old mice were transplanted into young mice. Remarkably, they regained their full regenerative ability. Likewise, when young systemic factors were introduced into aged animals—either through shared circulation or through plasma infusions—the old stem cells began to divide and repair as though time had been partially rewound.

These discoveries led scientists to hunt for the specific molecules in young blood responsible for this effect. One candidate, GDF11, showed promise in early studies. It appeared to enhance neurogenesis and improve vascular health in aged mice. Although subsequent replication has been mixed—some studies dispute its relevance—the search for rejuvenating "youth factors" continues. Whether it's GDF11 or another molecule, the broader lesson remains: the stem cell's environment can modulate its behavior profoundly.

And that's the hopeful twist. Stem cell exhaustion may not be a one-way street. It's not merely a tale of attrition but of inhibition. Many of our regenerative cells are not gone. They're waiting—for clarity, for cues, for conditions to change. This insight reorients our focus. Instead of simply trying to inject more stem cells into an aging body, maybe we should ask: what signals do these cells need to act again? Can we clean the fog from the niche? Can we retune the frequencies?

The implications ripple beyond muscle. Similar findings have emerged in neural stem cells, in the blood-forming niches of bone marrow, even in the lining of the intestine. Across systems, aging seems to be less about the disappearance of builders and more about their reluctance to pick up tools in a collapsing workshop. That means interventions—biological, pharmaceutical, behavioral—may still have the power to wake dormant potential. The cells are still there. They just need to be reminded why they matter.

As with all aging insights, this one carries a caveat. Not all silence is safe. Some stem cells are dormant for good reason. Pushing them to act without repairing the niche first could risk chaos—uncoordinated repair, fibrosis, even cancer. But with precision, with timing, with care, the idea that our regenerative decline is not destiny opens a new frontier.

The workshop is empty. But not abandoned. The tools are dusty. But not broken. And somewhere in the rafters, a clock still ticks—waiting for the bell that tells the apprentices it's time to return.

WHEN THE NICHE GOES ROGUE

In an overgrown garden behind a long-abandoned home, you can still find seeds. Under the mulch and tangled weeds, the promise of life lies dormant. But nothing grows. The sun still shines, rain still falls—but the soil is depleted, acidic, and littered with debris. Once, this soil nourished vibrant growth. Now it strangles it. The seeds didn't die. The garden did.

This is the story of a stem cell's microenvironment, its niche—and what happens when it breaks down with age.

The niche is more than a physical space. It's a biochemical ecosystem: rich with signaling molecules, extracellular matrix (ECM) scaffolds, immune modulators, oxygen gradients, and mechanical forces. It's where stem cells rest, renew, or leap into action when damage calls. Every stem cell type has its unique niche: hematopoietic stem cells reside in the bone marrow's spongy matrix; muscle satellite cells wedge beneath the basal lamina of myofibers; neural progenitors cluster in ventricular zones bathed in cerebrospinal fluid. Each is exquisitely tuned to its environment.

A healthy niche offers signals that say: "You're needed. Divide now." An aging niche, by contrast, often murmurs: "Stay quiet. It's not worth it."

As we age, these microenvironments degrade. They stiffen. They inflame. They suffocate the very cells they were built to support.

Let's start with structure. The extracellular matrix—the protein-rich scaffolding that houses and signals to stem cells—undergoes dramatic shifts. Collagen crosslinking increases, making the environment mechanically stiffer. Fibrotic remodeling thickens tissues with disorganized protein deposits, which distort cellular orientation and reduce nutrient diffusion. For muscle satellite cells, such stiffness alone can tip their fate—pushing them toward unwanted differentiation or quiescence. A stem cell that can't feel its environment properly often stops responding altogether.

Then there's the chemical landscape. Chronic, low-grade inflammation—sometimes called "inflammaging"—bathes the niche in cytokines like IL-6, TNF-alpha, and IFN-gamma. These inflammatory signals do more than annoy; they sabotage. In bone marrow, they shift hematopoietic stem cells toward myeloid bias, impair lymphoid regeneration, and increase the risk of malignancy. In muscle, they block satellite cell activation, turning once-vigorous repairmen into passive bystanders. And in the brain, they contribute to stem cell dysfunction that may underlie neurodegenerative drift.

Even oxygen stops flowing properly. Microvascular rarefaction—the loss of small capillaries—is common in aging tissues. With fewer vessels, oxygen tension rises or falls abnormally, destabilizing the delicate balance stem cells need to maintain proper metabolism. Many stem cells are metabolically finicky, thriving only in hypoxic niches with precise redox signaling. When blood flow fades, so does their identity.

And perhaps most subtly of all, the aged niche starts miscommunicating. The signaling pathways that once orchestrated renewal—Notch, Wnt, TGF-beta, IGF—are often misregulated. Aged muscle shows excessive TGF-beta activity, which induces fibrosis and represses satellite cell proliferation. In the aging brain, reduced Wnt signaling impairs neural progenitor expansion. These are not catastrophic mutations. They are drifted whispers—once-clear messages now garbled by time.

This isn't just a cell-intrinsic problem. The niche often fails first.

In one remarkable study, researchers used genetic tools to reprogram the niche itself—not the stem cells—restoring a youthful environment in aged muscle. The result? Dormant satellite cells woke up and began proliferating again. Other studies have used pharmacological agents to block niche-derived inhibitors like TGF-beta or activate dormant signaling pathways, with similar rejuvenating effects.

And yet, targeting the niche remains one of the most underexplored avenues in regenerative medicine. It's harder to reach, more complex, less genetically malleable. It's not one cell, but many—fibroblasts, immune cells, endothelial cells—all dancing to a deteriorating rhythm.

Still, the logic is compelling. If the garden has gone bad, don't blame the seeds until you've fixed the soil.

There's an even deeper implication here. If we can revive old stem cells by restoring their niche, then perhaps age itself is not wholly irreversible. Perhaps aging is more about signal loss and environmental decay than hardwired decline. That possibility isn't just scientifically provocative—it's profoundly hopeful.

Of course, this cuts both ways. Reviving the niche might reawaken not just healing but risk. An overly eager niche could drive stem cells into excessive proliferation, increasing the risk of cancer. The body has likely evolved to dial down niche signaling with age as a protective measure. Reversing that decision demands caution.

But the central insight holds: aging stem cells are shaped by their habitat. Fix the habitat, and the cell might not be "exhausted" at all.

Back in our garden, a new gardener arrives. They till the soil, prune the weeds, enrich the earth. And, eventually, the dormant seeds stir. Shoots emerge. Growth returns—not from new seeds, but from old ones given the right environment to thrive again.

That's the wager of niche rejuvenation. Not to replace what's lost, but to restore what's been silenced. And in doing so, reclaim the regenerative vitality that age so often seems to steal.

THE STEM CELL PARADOX

In 2006, a group of Japanese scientists did something that seemed to border on alchemy. They took adult skin cells—fibroblasts, the mundane foot soldiers of connective tissue—and with a cocktail of just four genes, reprogrammed them into something resembling embryonic stem cells. These were not just rejuvenated cells. They were reborn, capable of becoming almost any tissue in the body. They called them induced pluripotent stem cells. The lead researcher, Shinya Yamanaka, would later win the Nobel Prize for this work. But in that moment, he stumbled into a biological paradox that haunts regenerative medicine to this day: the same force that might one day help us reverse aging could just as easily unleash cancer.

Stem cells walk a fine line. By design, they are capable of extraordinary feats: dividing repeatedly, repairing damage, replacing the wornout. This potential comes from their ability to self-renew and to differentiate—to become specialized cells like neurons, muscle fibers, or immune warriors. But that same potential also makes them inherently dangerous. High proliferative capacity is a defining feature of cancer. In fact, some of the worst cancers—leukemias, glioblastomas—are driven by rogue cells that behave eerily like stem cells: fastdividing, self-sustaining, evasive.

From an evolutionary perspective, this danger isn't just theoretical. It's central to why stem cells might be silenced as we age. Antagonistic pleiotropy, the evolutionary concept that traits beneficial in youth can become harmful later, offers a compelling explanation. In youth, robust stem cell activity gives organisms a leg up in survival—healing wounds, fending off infections, building muscle and blood. But if those same cells remain hyperactive into later decades, they increase the risk of unchecked growth. So evolution makes a trade. It restrains the stem cells. Aging becomes the price of minimizing cancer.

This theory is supported by a pattern seen across species. Long-lived mammals, like whales and elephants, often have more tumor-suppressing genes or more conservative stem cell dynamics than their short-lived cousins. In mice, where lifespan is short and evolutionary pressure for latelife cancer control is minimal, stem cells remain more active later in life—at the cost of higher cancer rates. Nature seems to opt for survival through reproduction, not longevity. If dialing down regeneration after reproductive years prevents cancer, that's an acceptable cost to the species, if not the individual.

But what happens when we try to intervene?

In the wake of Yamanaka's breakthrough, researchers began experimenting with partial reprogramming—short bursts of reprogramming factor expression that might roll back the biological clock without fully wiping a cell's identity. In mice, these short pulses rejuvenated cells, improved tissue function, and in some studies, extended lifespan. But full reprogramming, or sustained exposure to the reprogramming genes, often led to tumors. Cells forgot who they were. They divided uncontrollably. The rejuvenation experiment turned into an oncogenesis experiment.

Even less radical interventions carry risk. Stimulating aged muscle stem cells to reactivate and rebuild tissue might improve healing—but if those cells have already accumulated mutations or epigenetic damage, revving their engines could lead to clonal expansion of faulty cells. Similarly, treatments that flush out inhibitors in the stem cell niche, like inflammatory cytokines, might unmask not just sleeping heroes but potential villains. A rejuvenated stem cell with a damaged genome is a loaded gun.

This is not just theoretical. Hematopoietic stem cells, which maintain the blood and immune system, often develop mutations with age—a phenomenon called clonal hematopoiesis. These mutated clones expand in the bone marrow, producing skewed blood profiles and raising the risk for blood cancers like acute myeloid leukemia. Imagine enhancing these cells' activity indiscriminately. You might amplify not just regeneration, but also malignancy.

Thus, precision becomes the operative word. We can't afford to rejuvenate blindly. The next phase of regenerative medicine must be about targeted activation—identifying stem cells that are genetically sound and selectively reawakening them while either avoiding or replacing the dangerous ones. This demands exquisite diagnostic tools, capable of parsing cellular age, mutation burden, and microenvironmental context.

One avenue being explored involves Yamanaka factor mimetics—compounds that tweak the same epigenetic pathways but with less force. Others focus on boosting mitochondrial health in stem cells, since energy deficits often correlate with stem cell dormancy. Still others look to senolytics—drugs that clear out senescent cells that might be suppressing stem cell function through inflammatory noise. Each of these strategies comes with its own set of risks and rewards.

The real danger may be hubris—our assumption that more regeneration is always better. The truth is, stem cells are gatekeepers of youth and agents of risk. They are the match and the fuse. Reigniting them is seductive, but without careful containment, we might just be lighting fires we can't put out.

So we're left with a delicate balancing act. How do we nudge the regenerative machinery without unleashing malignancy? How do we distinguish the cells that are "asleep" from those that are "corrupt"? How do we rejuvenate not just a body, but a system that maintains its own internal safety checks?

The future of aging medicine may hinge on these questions. And like Yamanaka in 2006, we may find that solving one problem opens the door to another. But if we learn to walk that tightrope—between regeneration and restraint, repair and risk—we might not just extend life, but improve the quality of the years we have.

Because aging, in the end, is not just the absence of youth. It's the accumulation of tradeoffs. And stem cell exhaustion is one of the sharpest—where the very tools of renewal become too dangerous to wield.

HOPE IN SUSPENSION

At dawn, the trauma ward's silence was broken only by the rhythmic beeping of monitors and the muted rustle of nurses' shoes. The patient—a woman in her seventies—had fractured her hip weeks ago. At first, the doctors weren't optimistic. Healing in someone her age, especially after a complex break, usually dragged on for months. She'd been immobile, uncomfortable, and emotionally drained. But something had shifted. Her followup X-rays now showed accelerated healing. She was regaining strength. The surprise wasn't in the recovery itself—it was in the cause.

She was part of a clinical trial testing a new type of plasma-based therapy. The premise was simple and radical: instead of replacing her depleted stem cells with someone else's, the goal was to clean the biochemical neighborhood those cells lived in. The logic was rooted in a strange but well-replicated observation from rodent studies: when the blood of young mice is introduced into old ones, the old ones heal faster. Their stem cells wake up. Their muscles regenerate. Their brains even spark new neuron growth. No gene therapy, no organ transplant—just a better environment.

The term for this field of therapy is systemic rejuvenation. It's less about injecting youth and more about removing—or diluting—the molecular noise that comes with age.

Consider this: as we grow older, our blood accumulates a cocktail of proinflammatory cytokines, senescenceassoci-ated secretions, and metabolic byproducts that act like static over a radio. Even if the body's internal repair teams—stem cells—are present, they can't hear the repair signals clearly. The message is lost in transmission.

One of the most direct efforts to address this has been therapeutic plasma exchange. The procedure is remarkably similar to dialysis: old plasma is drawn from the patient and replaced with a synthetic or donor alternative, aiming to flush out toxic factors. While initially controversial, early studies—both in mice and small human cohorts—suggest it can improve cognitive function and physical resilience. Importantly, it doesn't introduce anything new; it simply clears the clutter.

What's clearer, however, is that aging doesn't degrade all stem cells equally or irreversibly. In some tissues, the cells are still there but dormant, inhibited by chronic inflammation, fibrotic tissue, or dysfunctional support cells. The niche—the tiny microenvironment surrounding each stem cell—is often the real culprit. This has opened the door to therapies aimed not at the stem cells themselves, but at their habitats. mTOR inhibitors, like rapamycin, have emerged as one such tool. Originally developed as an immunosuppressant, rapamycin is now being explored for its anti-aging effects. In mice, it has extended lifespan and improved stem cell function, in part by reducing metabolic stress and inflammation in stem cell niches. By quieting the noise of aging, it may help old cells "hear" the cues to repair again.

Senolytics offer another complementary approach. These are drugs designed to eliminate senescent cells—those zombie-like cells that no longer divide but won't die either, and which release a toxic mix of inflammatory signals that poison their surroundings. By clearing out the dead weight, senolytics may rejuvenate tissue not by adding anything, but by subtracting what's harmful. Trials using combinations like dasatinib and quercetin have shown early promise, especially in diseases like idiopathic pulmonary fibrosis and osteoarthritis, where senescent cells play a clear role.

Taken together, these therapeutic frontiers suggest a new philosophy in regenerative medicine: reactivation, not just replacement. For years, the dominant vision was to inject youthful stem cells into aged bodies, hoping they'd rebuild from within. But results were mixed, and concerns about immune rejection and cancer risk loomed large. Now, researchers are increasingly asking whether we even need to transplant. What if the body has enough stem cells already—dormant, suppressed, but salvageable?

That debate—reactivation versus replacement—runs through the heart of modern aging research. Transplants remain essential in extreme cases, like patients whose bone marrow has been destroyed by chemotherapy. But for the broader challenge of age-related decline, many scientists believe we may get farther by tuning the environment than by swapping out the parts.

This new direction isn't without caution. Modifying the stem cell niche can have unpredictable consequences. Growth signals that awaken dormant cells might also promote oncogenesis. Dampening inflammation too aggressively could impair immune surveillance. Like any system with feedback loops, the body's repair machinery is delicately balanced. Nudge one dial too far, and the whole control panel might overload.

Still, the trauma patient's progress speaks for itself. Her

healing wasn't instant. There were still setbacks, still days of fatigue. But the arc bent differently—toward recovery, not stasis. And she wasn't alone. Around the world, labs and clinics are shifting their gaze not just toward prolonging life, but restoring vitality. Stem cell exhaustion, once seen as a biological culde-sac, is increasingly framed as a bottleneck that can be widened. Not effortlessly. Not universally. But possibly.

In a sense, the body is like a city whose construction crews are aging out of the workforce. The buildings are still sound. The tools are still in the shed. But the blueprints are fading, and the foremen are harder to reach. What if we could fix the phones? What if we could shine more light on the blueprints? The workers might still be there—waiting. Tired, maybe. But not finished.

The story ends not in miracle, but in motion. A woman walks again, carefully but steadily. A stem cell hears its cue. Somewhere in the bloodstream, a signal gets through. And for a moment—perhaps longer—the work of repair resumes.

Reader-tool boxes for this chapter

CLOSER LOOK ***Why stem cell exhaustion is downstream of every other hallmark***

Stem cells sit at the intersection of every other aging hallmark. Telomere attrition shortens their replicative reserve. Genomic instability erodes their fidelity. Epigenetic drift erodes their identity. Senescence claims those that get pushed too hard. Inflammaging changes the chemical climate of the niche around them. By the time you measure stem cell exhaustion in an old tissue, you are measuring the cumulative output of every other cellular failure mode upstream.

DePinho's group made this argument explicit in a 2010 Nature review, framing aging as a network problem in which telomere damage activates p53, which depletes stem cell pools and damages mitochondria — a cascade that radiates outward into tissue dysfunction. The implication is operational: if you want to rejuvenate a tissue, fixing the stem cells alone may not be enough. You also have to fix the soil they sit in.

Sahin & DePinho (2010). Linking functional decline of telomeres, mitochondria and stem cells during ageing. Nature, 464(7288), 520–528.

CLOSER LOOK ***Heterochronic parabiosis — the experiment that started the young-blood debate***

In 2005, Irina and Michael Conboy's team at Stanford performed a now-classic experiment: surgically connecting the circulatory systems of young and old mice (heterochronic parabiosis), then injuring the muscle of the old animal. The aged muscle stem cells, normally sluggish to respond, regenerated as if young. Notch signaling was restored. Tissue repair quality returned. The implication: old stem cells are not necessarily broken — many are fine, just sitting in a chemical environment that no longer instructs them properly.

The experiment launched a decade of work on circulating youth factors and their inverse, age-promoting factors. Some of that work has held up; some has not. GDF11, initially proposed as a youth factor, was later shown by independent labs to be more complicated than the early reports suggested. The lesson the field has settled on is more nuanced than the headlines: the niche matters as much as the cell, and rejuvenating either one without the other is unlikely to work.

Conboy et al. (2005). Rejuvenation of aged progenitor cells by exposure to a young systemic environment. Nature, 433(7027), 760–764.

EASY ACCESS ***Sarcopenia — the most visible stem cell story***

The muscles you have at seventy are not the muscles you had at thirty. In the average untrained adult, peak muscle mass arrives in the late twenties. By the seventh decade, total muscle mass has dropped by 30–40%. Some of that is disuse. Some is hormonal. But a substantial share is a stem cell problem — the satellite cells that flank every muscle fiber and step in to repair daily microdamage become fewer and slower with age. Resistance training keeps them in the game longer than anything else we have. It does not stop the clock. It changes the slope.

CLOSER LOOK ***The niche — why old stem cells aren't always broken***

Every stem cell lives inside a microenvironment called a niche — a specific arrangement of supporting cells, extracellular matrix, oxygen tension, and signaling molecules that tells the stem cell when to stay quiet, when to divide, and what to differentiate into. Aging changes every component of the niche. Supporting cells secrete more inflammatory cytokines. The matrix stiffens. Capillary networks thin. Wnt and Notch signaling drift.

The therapeutic implication is uncomfortable. Most early stem cell therapies tried to add young, healthy cells back into old tissue — and watched the transplanted cells age within the recipient's environment. More recent strategies focus on rejuvenating the niche itself: clearing senescent supporting cells, restoring matrix mechanics, normalizing inflammatory signals. The cell, it turns out, is rarely the limiting factor. The neighborhood is.

Brack & Rando (2008). Aging of the skeletal muscle stem cell niche. Methods in Enzymology, 449, 33–61.

KEY TERMS *Vocabulary of the stem cell chapter*

Adult stem cell — A tissue-resident cell that can self-renew and differentiate into the cell types of its tissue.

Quiescence — The reversible non-dividing state most adult stem cells spend most of their time in.

Niche — The microenvironment surrounding a stem cell, providing the signals that govern its behavior.

Satellite cell — The muscle stem cell. Sits between the muscle fiber and its basement membrane; activates after injury.

Heterochronic parabiosis — Surgical joining of the circulatory systems of a young and an old animal. The classic experiment for testing systemic youth factors.

BROKEN COMMUNICATIONS

INFLAMMAGING, THE IMMUNE SYSTEM, AND THE MICROBIOME

In the early 1960s, a social psychologist named Stanley Milgram conducted a curious experiment in the city of Wichita, Kansas. He mailed out hundreds of envelopes to randomly selected people, each containing a letter that needed to make its way to a target individual—someone who lived far across the country. The rules were simple: each participant could only send the letter to someone they knew personally, someone they thought might be closer to the final recipient. What fascinated Milgram was how quickly the letters often arrived—through a short chain of friends and acquaintances. He called it "the smallworld problem," later popularized as "six degrees of separation." What Milgram stumbled upon was the elegance of a well-connected network, one where even faraway nodes could communicate through surprisingly few steps.

But what happens when the network falters?

Imagine if those letters started getting delayed. Or routed to the wrong people. Imagine if some addresses had changed without notice, or if certain post offices now flagged every message as suspicious and quarantined it. The elegant, efficient web would devolve into noise—connections would fray, signals would garble, and what was once a graceful system would begin to stumble.

This is the story of aging—not as a collapse, but as a miscommunication. Not as sudden disease, but as a subtle cacophony.

By the time she turned eightyfive, Eleanor didn't complain of any particular illness. Her heart was fine. Her kidneys showed no signs of failure. Her bones had held up. And yet, she was tired in a way that wasn't explained by any lab test. Mornings began with a slow ache that never quite went away. She used to walk to the community mailbox—now even the hallway seemed long. Her mind felt foggy. She wasn't depressed, she insisted. Just\... worn out.

Her doctors ran blood work. The results showed something increasingly common in people her age: elevated inflammatory markers. Her levels of interleukin-6 (IL-6), interleukin-1β (IL-1β), tumor necrosis factor alpha (TNF-α), and C-reactive protein (CRP) were all modestly above normal. Not alarming enough to suggest infection or disease. But collectively, they painted a picture of chronic, systemic immune activation.

A condition that had once gone unnoticed for decades.

Now, it had a name.

Inflammaging.

Coined by Italian immunologist Claudio Franceschi in the early 2000s, inflammaging describes a phenomenon that is as widespread as it is invisible: a gradual, persistent rise in inflammatory signals in the body with age. Unlike the sharp spike of a fever, this is a low simmer—barely perceptible but everpresent. And unlike acute inflammation, which is protective and temporary, inflammaging tends to linger indefinitely.

It is not one thing. It is not caused by one trigger. It is a drift.

Aging bodies accumulate microdamage. Cells die (...or don't, as is the case of senescence). Debris builds up. The immune system, once adept at cleaning up, begins to misfire. It continues to send out alerts, even when the fire is out. It issues signals to repair, but fewer workers respond. Some of the alarms are set off by senescent cells—old, damaged cells that no longer divide but refuse to die. These cells release a cocktail of inflammatory molecules called the senescence-associated secretory phenotype (SASP), including many of the same molecules that showed up in Eleanor's blood work: IL-6, IL-1β, TNF-α.

Other alarms come from metabolic changes. Aging fat cells, especially those in the abdomen, secrete inflammatory molecules known as adipokines. Add to this a lifetime of low-level infections—bacteria in the gums, viruses that never fully left the system, occasional breaches in the gut lining—and the immune system is constantly nudged into action. Not full war. But a state of armed readiness that never quite winds down.

This isn't just biochemical trivia. Chronic inflammation seeps into every tissue.

In the brain, it accelerates the clumping of proteins associated with Alzheimer's disease. In the arteries, it fosters the development of plaques. In the muscles, it contributes to wasting. In the bones, it weakens the scaffolding. It is a ghost conductor—one that subtly derails every performance without being seen.

Franceschi's original papers noted that centenarians—people who live past 100—tended to have either low levels of inflammation or a curious balance: they had elevated proinflammatory markers but also elevated anti-inflamma-tory ones, like IL-10. It was not the absence of immune activity that protected them, but the harmony of the signals. Like a neighborhood where people still shout occasionally, but someone is always listening. The feedback loops still worked.

For most of us, that balance breaks down by our seventies or eighties. What begins as a whisper becomes a back-ground hum, and then a persistent buzz. For Eleanor, there was no disease to name. But she was living in a body where the system had forgotten how to rest. Where immune cells kept patrolling even in the absence of danger. Where signals meant to be acute had become chronic.

And like any neighborhood with too many sirens, too many false alarms, the people began to tune them out.

That, too, is a danger. Over time, the body becomes less sensitive to genuine threats—just as real emergencies can be ignored in a city flooded with noise. The immune system's vigilance becomes blunted by its own hyperactivity.

This is the paradox of inflammaging. It exhausts us not with fire, but with smoke. It wears us down not through catastrophe, but through constancy.

In aging, the problem is not that the orchestra stops. It's that the instruments play louder, out of rhythm, or with notes that no longer match the score. The immune system, once a master communicator, becomes an unreliable messenger. And what that produces is not disease in the traditional sense—but a state of being unwell.

Inflammaging is the sound of the city humming all night. Never quite letting the system sleep.

And unless it is re-tuned—unless the noise is modulated—aging becomes not just a biological process, but an internal dissonance we can't quite explain.

THE MISDIRECTED DEFENDERS

In the 1970s, researchers studying Antarctic expeditions stumbled upon a strange paradox. The extreme isolation and sterile environment meant these explorers encountered almost no new pathogens for months. Yet, upon returning home, many of them fell severely ill with viruses that should have caused only mild symptoms. It was as if their immune systems had forgotten how to respond properly. Over time, it became clear that immune function wasn't a static property but something deeply sensitive to context—environmental, psychological, and biological. Like a neighborhood watch that grew rusty in a peaceful season, the immune system could either overreact or go silent when finally challenged again.

Eleanor's immune system had never been to Antarctica. But the confusion it now exhibited was no less dramatic.

She stood at the sink, toothbrush in hand, halfwatching the late news on her kitchen TV. As she spat, a ribbon of pink swirled down the drain. Her gums bled slightly again—this had become more frequent lately. Her dentist had said it was mild gingivitis, nothing alarming. But she'd noticed something odd: ever since this started, she felt drained the next day. Sluggish. Not just from poor sleep, but as if her body was mounting a defense operation for a war that didn't exist.

Inside Eleanor, the skirmish was already underway.

The aging immune system is not a battlefield emptied of troops. It is more like a battalion with poor leadership, frayed communications, and faulty targeting systems. Its frontline defenders—macrophages, neutrophils, natural killer cells—remain present but prone to misfires. Its specialized cells—T and B lymphocytes that once operated with surgical precision—become fewer, slower, and less responsive with age.

This dual drift has a name: immunosenescence. And it isn't simply a weakening. It's a distortion.

Eleanor's bleeding gums had triggered a classic inflammatory cascade. In younger years, resident macrophages would have quietly surveyed the area, recruited neutrophils if needed, and signaled resolution within hours. But in an aged body, macrophages become less discerning and more reactive. They overproduce inflammatory cytokines. Their ability to clear pathogens or cellular debris—called phagocytosis—is impaired. The result is a clumsy overreaction, like security guards flooding the scene, tripping over each other, unable to distinguish friend from foe.

Meanwhile, the adaptive branch of the immune system, which should have tempered this response, was under-staffed. Eleanor's T cells, especially the naïve ones responsible for recognizing new threats, had dwindled. Years of exposure to viruses like cytomegalovirus (CMV) had consumed much of her T-cell diversity. Her body had become filled with old veterans—memory T cells trained for specific past battles—but had few recruits left for anything new. The B cells, too, were sluggish, producing fewer and lowerquality antibodies.

This misalignment between the two arms of immunity—hyperactive innate cells and dwindling adaptive ones—explains many of aging's immunological ironies.

Older adults get sicker from new infections, like influenza or COVID-19, not just because they're frail, but because their immune system doesn't rise to the occasion. Vaccines, which rely on robust T and B cell cooperation, often elicit weak or short-lived responses. The immune orchestra plays, but the melody is off-key, and key instruments are missing. This is why flu vaccines are reformulated in higher doses for seniors, and why novel vaccine strategies are being explored for the aging population.

At the same time, this weakened surveillance opens the door for old enemies to resurface.

Take shingles, for example—a painful rash caused by reactivation of the varicella-zoster virus, the same virus that causes chickenpox. After the initial infection, this virus goes dormant in nerve cells. In healthy adults, T cells keep it in check indefinitely. But with immunosenescence, those T cells lose grip. What was once a latent standoff becomes a fullblown siege. Eleanor had suffered a bout of shingles at 78, and the pain had lingered for months, far longer than the rash itself. It had felt like her nerves were remembering the war long after the fighting stopped.

The same pattern plays out with other chronic infections. Epstein-Barr virus, herpes simplex, and CMV persist in low-level form for decades, subtly activating the immune system, contributing to inflammaging, and draining resources from other responses. These are the biochemical equivalents of background noise—interference that keeps the immune network on edge.

Yet in this paradox lies another: while the system becomes less able to combat real threats, it becomes more likely to turn against itself.

Autoimmunity increases with age. Conditions like rheumatoid arthritis, polymyalgia rheumatica, and giant cell arteritis are more common in older adults. Misguided immune cells, perhaps in part due to epigenetic drift or chronic inflammation, begin to target the body's own tissues. Tolerance breaks down. Signals are misread. It's as if the immune system, losing the clarity of its youth, starts suspecting everyone—including its own citizens.

Eleanor had tested negative for most autoimmune diseases. But her chronic low-grade inflammation—her sore knees, her tiredness, her persistently high CRP—were whispers of this loss of immune fidelity. Her defenders were still at their posts. But they no longer waited for orders. They reacted first and asked questions later.

In a younger body, inflammation is a focused fire. It rises fast, burns hot, and then retreats. In an aging body, it becomes a smolder—too weak to destroy, too persistent to ignore. This is why inflammaging often feels like malaise. There is no fever, no clear infection. Just a creeping sense that something is off, like smoke leaking from under a closed door.

The metaphor that aging is an immune "decline" misses the complexity of the change. It is not just a matter of fewer soldiers, but of mixed allegiances and erratic orders. The system that once operated with confidence now behaves like a jittery police force—paranoid, overstaffed in the wrong places, understaffed where it matters, and increasingly prone to friendly fire.

This is what inflammation becomes in old age—not a calibrated response, but a broadcast of confusion. And confusion, in a system meant to protect, is its own kind of threat.

GUT INSTINCTS GO AWRY

In the late 19th century, Russian scientist and Nobel laureate Élie Metchnikoff noticed something peculiar about the Bulgarian villagers. They lived longer than expected, remained physically active deep into old age, and—most curious of all—drank a fermented milk drink religiously. Intrigued, Metchnikoff proposed that the secret to longevity lay not in their hearts or brains, but in their guts. He believed that certain intestinal bacteria slowed aging, and even suggested that consuming "beneficial microbes" could delay senility. At the time, his theory was ridiculed by many in the medical establishment. But over a century later, the spotlight has swung back—brighter and more focused than ever—on the gut as one of the most powerful control centers in the aging process.

The human gut contains trillions of microbes—more cells than there are stars in the Milky Way. When Eleanor was young, this population was richly diverse. There were fiberloving Bifidobacteria, short-chain-fatty-acid-producing Firmicutes, and a well-balanced assortment of other players, all contributing to digestion, immune education, and nutrient absorption. These microbes weren't just passive squatters. They talked constantly with her immune system—whispering, warning, soothing. In fact, up to 70% of immune cells reside near the gut, constantly surveying microbial chatter.

But with age, this inner city started to lose its symphony. One of the first casualties was diversity. The aging microbiome shrinks in complexity. Protective species like Bifidobacteria decline, while opportunistic strains—often linked to inflammation—begin to dominate. This microbial reshuffling isn't just cosmetic. It affects the kinds of molecules being produced, the way the gut lining is maintained, and even the mood-regulating signals being sent to the brain.

Intestines, like many in their 80s, also grown more permeable. This phenomenon, often called "leaky gut," involves the breakdown of tight junctions in the gut lining. These junctions are like customs officers, deciding which substances can cross into the bloodstream. But with age, stress, poor diet, and a degraded microbiome, they grow lax. Molecules that were never meant to circulate—like bacterial fragments, toxins, and metabolites—slip through.

One of the most notorious intruders is lipopolysaccharide (LPS), a molecule found on the surface of gramnega-tive bacteria. When LPS leaks into the bloodstream, it triggers a strong immune response. The body thinks it's under microbial attack. The result is a chronic, low-grade inflammatory reaction. One study found that older adults with high blood levels of LPS had elevated C-reactive protein and interleukin-6—classic markers of inflammaging.

In essence, Eleanor's body was waging a slow war on microbial ghosts. The gut was not under attack—but her immune system believed otherwise.

Science has begun to map this terrain more clearly. In one landmark experiment, old mice received fecal transplants from young mice. The results were stunning: inflammatory markers dropped, cognition improved, and even physical endurance increased. In other words, simply shifting the microbiome seemed to rewire the internal communication networks—restoring calm to a body that had grown confused.

The inverse was also true. Young mice given the microbiota of older mice showed accelerated signs of aging: poorer memory, increased inflammation, and reduced neurogenesis. The gut, it seemed, was not just reflecting the body's age. It was actively enforcing it.

But how? Part of the answer lies in the gutimmune axis—a bidirectional communication system connecting the digestive tract to the immune system, and through it, nearly every organ in the body. Microbial metabolites (like short-chain fatty acids, especially butyrate) can calm inflammation and promote gut integrity. Others, like TMAO, may do the opposite—fueling cardiovascular disease. With a balanced microbiome, the good outweighs the bad. With age, this balance tilts, and the immune system begins to respond to internal signals as if they were threats.

This microbial miscommunication also affects the nervous system. Elders often wake with foggy thinking and mild headaches—symptoms easily dismissed as "normal aging." But research has shown that microbiome-derived molecules cross the bloodbrain barrier, influencing neuroinflammation, mood, and cognitive resilience. Inflammation in the gut can stoke inflammation in the brain—a phenomenon now being studied in Alzheimer's research. What happens in the gut doesn't stay there. It echoes.

When Metchnikoff proposed that bacteria in yogurt could extend life, he was mocked for his simplicity. But his core idea—that the gut is not a passive pipe but a central player in health and aging—has now been validated. Eleanor's gut was not failing in isolation. It was a hub whose dysfunction rippled outward, disrupting immune calibration, inflammatory tone, and even emotional equilibrium.

The tragedy of aging, in some ways, is not that the machinery breaks, but that the conversations between parts of the system grow garbled. The microbiome is one of the most eloquent speakers in this internal dialogue—and one of the first to fall silent with time.

And yet, this silence is not irreversible.

Hope, it turns out, can be measured in bacteria per gram. In the whispers of the microbiome, there is a path to clearer signals, quieter alarms, and a body that once again recognizes what is friend and what is foe.

THE BURN OF HIDDEN FIRES

In the early 1990s, forest rangers in Yellowstone noticed a peculiar pattern: even after suppressing all visible fires for decades, trees were still dying. Roots rotted underground. Saplings failed to grow. Eventually, they found the culprit—not roaring flames, but a slow underground smolder in the peat. Invisible to the eye, yet persistent enough to alter the ecology of an entire landscape.

That is what chronic inflammation looks like in the aging body. Not a blaze, not an infection, not an emergency. Just a continuous simmering deep within the tissues, unnoticed and unalarmed, yet steadily eating away at the architecture of life.

Our 85-year-old protagonist gets her latest lab results: CRP, IL-6, TNF-α—markedly elevated. She isn't sick in the conventional sense. No fever. No pain. No infection. But her doctor frowns. "Your body," he says gently, "is fighting something. It just doesn't know what."

The concept of "silent inflammation"—or inflammaging — rests on the recognition that the immune system, once the great sentinel of the body, doesn't always shut off its alarms in old age. Even in the absence of external threats, it continues to patrol, to signal, to agitate. But the causes of that agitation are not pathogens—they are the debris of aging itself.

One major culprit is the senescent cell. As explored in a previous chapter, these are cells that have exited the cell cycle permanently but refuse to die. Instead, they hang around secreting a cocktail of inflammatory molecules—a phenomenon known as the Senescence-Associated Secretory Phenotype (SASP). Like elderly neighbors shouting gossip across balconies all day, they don't contribute meaningfully anymore—but they sure make noise. And that noise reaches far beyond their immediate surroundings. TNF-α, IL-1β, MMPs, reactive oxygen species—they leak out into the tissue, priming immune cells into a state of constant vigilance.

Then there's visceral fat—the deep belly fat that accumulates with age. More than just passive storage, it's biologically active tissue. Adipocytes in this fat depot begin releasing adipokines—cytokines with inflammatory properties. Leptin levels spike. MCP-1 (monocyte chemoattractant protein) draws in more immune cells. The fat depot becomes a hub of inflammation, feeding the fire from the center of the body outward. That's why waist circumference isn't just a cosmetic issue in aging—it's a metabolic and immunological one.

Add to this the invisible enemies: chronic viral infections. Cytomegalovirus (CMV), for example, can persist in the body silently for decades. It doesn't cause symptoms, but it keeps the immune system slightly activated—especially in aging adults with already-compromised immune surveillance. The result? A background hum of interferons and inflammatory cytokines that never really turns off.

Another subtle provocateur is damage itself. The very act of aging produces byproducts—misfolded proteins, oxidized lipids, broken DNA fragments. These are known as Damage-Associated Molecular Patterns (DAMPs), and they're picked up by the immune system's pattern recognition receptors. Toll-like receptors, NOD-like receptors, and inflammasomes respond as if they've encountered a pathogen. But there is no pathogen. Only the cellular ruins of time. And so, the immune system launches a low-grade, misdirected response against its own degeneration.

This low-grade inflammation doesn't just coexist with aging—it accelerates it. Chronic inflammation shortens telomeres by increasing cellular turnover and oxidative stress. It drives epigenetic drift by modifying the expression of genes involved in immune regulation. It contributes to sarcopenia by interfering with anabolic signals in muscle. It even impairs stem cell function by altering their niches, as explored in the previous chapter.

Worse still, inflammation begets more inflammation. The immune system's attempt to clear senescent cells may backfire, causing collateral tissue damage and releasing more DAMPs. Inflammatory cytokines increase vascular permeability, which lets more microbial fragments into circulation from the gut. And as tissues become damaged, they leak signals that draw more immune cells into a futile repair effort. This is the paradox: the very system designed to maintain order starts to break it down.

So, what does it mean when a lab test shows "mildly elevated inflammation" in an elderly person with no clear illness? It means the smoldering has begun. It means the forest may not be on fire—but the roots are heating up. And while we can't always hear it, the burn continues in silence, reshaping the landscape of the body one damaged cell at a time.

NOT ALL INFLAMMATION IS THE ENEMY

At a crowded international symposium on aging biology, the lights dim and the immunologist takes the stage. No slides yet. Just a story.

"When I was a young researcher," he begins, "I thought inflammation was the villain. I imagined the immune system as a fire department, and aging as a case of broken sprinklers—water everywhere, flooding the place, damaging the floors. I spent two decades trying to shut the valves. Then I visited Okinawa."

He clicks, and the slide changes to a photo: a smiling 103-year-old woman in a sunhat, planting sweet potatoes. "Her bloodwork stunned me," he says. "Elevated IL-6, TNF-alpha, even some CRP. But also—skyhigh IL-10 and TGF-beta. Her immune system wasn't silent. It was singing in harmony."

A pause.

"We've been trying to silence a choir when we should have been rewriting the music."

For years, the narrative around aging and inflammation has been built on the assumption that less is better. And understandably so—chronic inflammation correlates with nearly every age-related disease: atherosclerosis, Alzheimer's, frailty, insulin resistance, even certain cancers. The idea that the aging immune system resembles a dysfunctional smoke alarm—blaring at shadows—has driven an entire generation of anti-inflammatory interventions.

But what if the volume isn't the only problem? What if the composition matters more?

The emerging field of "successful inflammaging" proposes exactly that. First coined in studies of centenarians in Italy and Japan, the term refers to individuals who show elevated proinflammatory markers—yes—but also unusually high levels of anti-inflammatory mediators. It's not a matter of absence, but balance. Like a thermostat that still works, even if the room has warmed. The system hasn't shut down; it's recalibrated.

Take IL-6, often cited as a prime villain in chronic inflammation. High levels of IL-6 predict mortality, cardiovascular disease, and cognitive decline. But IL-6 also plays a critical role in muscle regeneration, liver repair, and metabolic homeostasis. During exercise, muscle fibers release IL-6 as a myokine—a beneficial signal that mobilizes energy and immune coordination. To treat all IL-6 as pathological is like treating every raised voice as a threat. Context is everything.

This nuance is where many pharmaceutical efforts have stumbled. Consider the CANTOS trial—one of the most ambitious anti-inflammatory studies ever conducted. In it, researchers administered Canakinumab, a monoclonal antibody targeting IL-1β, to over 10,000 patients with prior heart attacks and elevated CRP. The results? A 15% reduction in recurrent cardiovascular events. Proof that targeted inflammation reduction can work.

But the trial had a twist: patients also experienced significantly more infections, some fatal. The immune system's dual role—as both firefighter and gardener—means that dampening one function can cripple another. The price of turning down inflammation isn't always visible on a blood test—it might be paid at the microbial frontlines.

Other interventions targeting TNF-alpha or JAK-STAT pathways have shown similar dilemmas: yes, they reduce markers of inflammation. But they also suppress wound healing, immune surveillance, and even tissue regeneration. What began as elegant solutions have, in many cases, morphed into blunt instruments.

Meanwhile, in realworld longevity, the picture grows more intriguing. Studies of centenarians have uncovered something paradoxical: many of them exhibit elevated inflammatory markers compared to younger adults—but without the usual disease burden.

A Japanese study of over 200 centenarians found that while IL-6 and CRP were indeed higher than in middleaged controls, these elders also had elevated levels of IL-10, an anti-inflammatory cytokine, and higher ratios of T-regulatory cells. This suggests a compensatory mechanism—akin to a seesaw that may tilt, but never topples.

These individuals also tend to have more diverse T-cell repertoires and better innateadaptive cross-talk. Their immune systems, though aged, are symphonic. Slightly raspy in tone, perhaps—but still in tune.

Researchers now speculate that longevity may depend less on suppressing inflammation and more on maintaining immune resilience. That means fostering systems that can rise to challenges but return to baseline—a flexibility that aging often compromises. Like a neighborhood siren that sounds in emergencies, but also knows when to turn off.

This shift in framing has implications beyond research. It changes how we think about interventions. The question is no longer "How do we stop inflammation?" but "How do we regulate it wisely?"

It means:

Promoting immune literacy—not just dampening symptoms.

Encouraging lifestyle changes that gently re-tune inflammation (e.g., regular exercise, sleep, and polyphenol-rich diets).

Developing therapeutics that modulate, rather than silence—such as immune checkpoint adjusters or metabolic rebalancers.

Investigating microbiome interventions, which may shift inflammatory tone from the bottom up.

In one recent study, caloric restriction in elderly individuals reduced inflammatory markers without compromising immune function. Another found that exposure to novel microbial environments—think forest air, fermented foods—can recalibrate immune tone in just weeks. These interventions don't block inflammation; they retune it, like adjusting the strings on a violin.

REBOOTING THE SYMPHONY

In 1967, a young conductor named Seiji Ozawa stood before the Berlin Philharmonic—one of the most formidable orchestras in the world. His audition didn't begin with a rousing overture or a grand flourish. Instead, he started with silence. He simply stood there, waiting. Musicians fidgeted. Violin bows hovered. And then, with the gentlest wave, he cued a single oboe note.

It wasn't about power—it was about trust.

Within minutes, he'd reoriented the entire orchestra, not by commanding, but by listening. He tuned the players to each other. Not a single section was replaced. Yet by the end, they played as if reborn.

This is what rebooting the aging immune system may look like. Not a war against decline, but a subtle re-coordination of parts that still remember how to work together—if only given the right cue.

The emerging field of rejuvenation medicine is pivoting on exactly this principle: restoring communication. Not erasing age, but tuning function. And the interventions that show promise aren't necessarily exotic or futuristic—they're often rooted in ecology. Because the body isn't a machine—it's a living ecosystem, and ecosystems can recover when their channels reconnect.

Let's explore the most exciting—and diverse—levers of this biological reboot.

Senolytics

The first approach is subtractive. Imagine an orchestra, maybe, where several musicians have stopped playing—but won't leave the stage. Worse, they complain constantly during the performance, disrupting others.

Senolytic therapies aim to remove these cells selectively. Compounds like Dasatinib and Quercetin have shown promise in preclinical trials—clearing senescent cells from aged tissues and reducing inflammatory load. In mouse models, this led to improved cardiovascular function, enhanced muscle regeneration, even delayed onset of frailty.

But timing is everything. Kill senescent cells too early, and you lose their beneficial roles in wound healing and cancer suppression. Like pruning a tree, precision is key.

Rewilding the Gut

Parallel to this is an additive strategy: changing the community inside us.

Microbiome transplants—especially from young donors to old recipients—have gained traction. In rodents, fecal microbiota transplantation (FMT) has reversed age-related memory decline, improved immune tone, and reduced systemic inflammation. Aged mice receiving gut flora from young counterparts regained exploratory behavior and showed rejuvenated hippocampal gene expression.

But transplants are only one piece. Prebiotics—indigestible fibers that feed beneficial bacteria—can shift the microbial balance toward anti-inflammatory species like Bifidobacteria and Faecalibacterium. Postbiotics, such as butyrate or other bacterial metabolites, offer a shortcut: delivering the microbial benefits without needing the microbes themselves.

What matters isn't just the bacterial species—it's the messages they send. And those messages can be rewritten with food, environment, and supplementation.

Nutrient-Sensing Modulation

The immune system doesn't act in isolation. It follows cues from metabolic pathways—particularly those governed by nutrient sensors like mTOR, AMPK, and SIRT1.

Rapamycin, an mTOR inhibitor, has emerged as a frontrunner in longevity studies. By dampening mTOR signaling, it mimics caloric restriction—an intervention long known to extend lifespan in multiple species. In aged animals, rapamycin improves vaccine responses and reduces inflammatory markers. Human trials are underway, cautiously exploring optimal dosing and safety.

Meanwhile, metformin, originally a diabetes drug, modulates AMPK and may reduce inflammatory tone. Resveratrol and NAD+ boosters influence sirtuin activity, intersecting with DNA repair and mitochondrial communication.

Together, these compounds represent a shift from symptom suppression to signal tuning—targeting the upstream regulators of intercellular dialogue.

Young Plasma and the Dilution Hypothesis

Among the more surprising rejuvenation strategies is not what you add—but what you remove.

Studies by researchers like Tony Wyss-Coray have shown that diluting old plasma—even without adding anything young—can improve tissue function in aged mice. The implication? Aging isn't just the absence of youth factors, but the presence of active inhibitors: inflammatory cytokines, degraded proteins, and possibly microRNAs that reinforce dysfunction.

By filtering the blood and replenishing it with albumin-rich saline, scientists saw restored neurogenesis and muscle regeneration. This opens a new paradigm: that the bloodstream is not merely a carrier, but a regulatory environment, capable of suppressing or enabling repair.

Human trials using plasma dilution and exchange are underway.

Hypothalamic Inflammation as the Master Switch

One of the boldest ideas in aging science is that the brain sets the pace—and particularly, a tiny region called the hypothalamus.

In a landmark study, researchers found that reducing hypothalamic inflammation in mice—via inhibiting NF-kB signaling—led to longer lifespans, better muscle function, and improved metabolic profiles. Conversely, inducing inflammation in the hypothalamus accelerated aging hallmarks across tissues.

Why? The hypothalamus integrates hormonal, neural, and immune signals. It's the body's master conductor. Inflammaging, then, may begin not in the periphery, but in the central control room. If true, brain-targeted therapies—ranging from microglial modulators to circadian rhythm restoration—could be keys to systemic rejuvenation.

Extracellular Vesicles

The final frontier lies in what cells say to each other—not via hormones or cytokines, but through extracellular vesicles (EVs). These tiny packages of RNA, proteins, and lipids act as targeted messages, altering gene expression in recipient cells.

In young animals, EVs promote regeneration and suppress inflammation. In aged individuals, their content shifts—

sometimes promoting senescence, fibrosis, or immune dysfunction. Modulating EV composition—either through exercise, diet, or engineered vesicles—is now being explored as a way to retrain intercellular communication.

Imagine giving the cells new scripts—so they remember how to cooperate.

Biologically, perhaps not much has changed. But the conversations within have. And that, more than any miracle drug, is the real measure of progress. Because aging isn't the collapse of machinery—it's the misfire of coordination. And the future of longevity may lie not in replacing parts, but in re-tuning the players.

Just like Seiji Ozawa before the Berlin Philharmonic, what the aging body needs isn't louder instruments or flashier solos—but a steady hand, a gentle cue, and the belief that the music can still be made whole.

Reader-tool boxes for this chapter

> CLOSER LOOK *Inflammaging — the term, the mechanism, the stakes*

The term inflammaging was coined by Claudio Franceschi and colleagues in 2000 to describe a low-grade, chronic, systemic inflammatory state that develops with age in the absence of overt infection. It is detectable as modest elevations in IL-6, CRP, and TNF-α — small numbers that, sustained over decades, correlate strongly with frailty, cardiovascular disease, dementia, and all-cause mortality.

A 2019 Nature Medicine review by Furman and colleagues catalogued the breadth of evidence linking systemic chronic inflammation to nearly every major non-communicable disease — cardiovascular, metabolic, neurodegenerative, autoimmune, and oncologic. The picture that emerges is not of inflammation as a side effect of aging but of inflammation as a central node in the aging network, with inputs from senescent cells, gut barrier dysfunction, and immunosenescence converging into the same persistent signal.

Furman et al. (2019). Chronic inflammation in the etiology of disease across the life span. Nature Medicine, 25(12), 1822–1832.

EASY ACCESS ***Your microbiome, in numbers***

The reference 70-kg human carries roughly 38 trillion bacterial cells, mostly in the colon, weighing about 200 grams together — roughly the mass of a hamster, distributed through the gut. The widely repeated 'we are

10% human' figure was a 1970s estimate based on rough arithmetic; modern measurements put the bacterial-to-human cell ratio closer to 1.3 to 1. That is still extraordinary. Functionally, your microbiome encodes about 100 times more genes than your own genome, contributing to digestion, immune training, neurotransmitter synthesis, and barrier maintenance. When that community shifts with age — fewer Bifidobacteria, more pro-inflammatory species, less diversity overall — the consequences travel well outside the gut.

CLOSER LOOK *The hypothalamic inflammation hypothesis*

In 2013, Dongsheng Cai's group at Albert Einstein published one of the most provocative papers in modern aging biology. They reported that NF-κB activation in the hypothalamus — the deepest, oldest part of the brain, governing hormones and homeostasis — increases with age across rodent models, and that genetically blocking that activation (or surgically removing the IKKβ pathway in hypothalamic neurons) extended lifespan and slowed multiple aging phenotypes.

The proposed mechanism is striking: hypothalamic inflammation suppresses gonadotropin-releasing hormone (GnRH) production, which in turn impairs adult neurogenesis and propagates aging signals to the periphery. GnRH replacement partially rescued the phenotype. Whether the same circuitry operates in humans remains under active investigation, but the paper introduced a powerful idea — that aging may have a central command node, and that node may be tractable.

Zhang et al. (2013). Hypothalamic programming of systemic ageing involving IKK-β, NF-κB and GnRH. Nature, 497(7448), 211–216.

CLOSER LOOK ***CANTOS — when an anti-inflammatory drug worked***

The CANTOS trial enrolled over 10,000 patients who had survived a heart attack and had elevated CRP. They were randomized to canakinumab — a monoclonal antibody that neutralizes interleukin-1β, a key inflammatory cytokine — or placebo. After roughly four years, canakinumab significantly reduced the rate of recurrent cardiovascular events, with the effect concentrated in patients whose CRP fell most. The drug also unexpectedly reduced lung cancer incidence and mortality.

CANTOS was not a longevity trial. It was a cardiology trial. But it answered a long-standing question: if you turn down inflammation in people, without lowering cholesterol or any other classical risk factor, do they fare better? The answer was yes, with caveats — the drug increased fatal infections, the effect required ongoing administration, and the cost was high. Still, the proof of principle was clean. Inflammation is not just a marker. It is a driver.

Ridker et al. (2017). Antiinflammatory therapy with canakinumab for atherosclerotic disease. New England Journal of Medicine, 377(12), 1119–1131.

KEY TERMS ***Vocabulary of the inflammaging chapter***

Inflammaging — Low-grade, chronic, sterile inflammation that develops with age. Term coined by Franceschi in 2000.

Immunosenescence — Age-related decline in adaptive and innate immune function. Reduces vaccine response and pathogen clearance.

Microbiome — The community of microorganisms (mostly bacteria) living on and inside us. The gut population is the largest by far.

NF-κB — A central transcription factor in inflammatory signaling. Persistently elevated in many aging tissues.

CRP — C-reactive protein. A standard clinical readout of systemic inflammation. Elevated levels predict cardiovascular risk.

Dysbiosis — A shift in microbial community composition associated with disease. Common with aging and antibiotic use.

THE BIG PICTURE

INTEGRATING THE HALLMARKS

There's a city built along a river. For decades, it thrived. Not because it was perfect, but because it worked in sync. Power stations supplied electricity precisely when homes and factories needed it. Garbage trucks ran every morning, clearing the debris before it piled. Water filtration plants balanced the river's offerings—sometimes muddy, sometimes clean—into something drinkable. The mayor never had to worry about the lights staying on or traffic choking intersections. Even the graffiti artists and latenight jazz clubs had their place.

But then, slowly, without a storm or war or collapse, something changed. The traffic lights grew sluggish, blinking out of rhythm. Garbage pickups missed entire neighborhoods. Power surged in some places and vanished in others. No building fell, no explosion marked the day things turned—but the city started wheezing. Not from catastrophe. From dissonance. Its subsystems no longer spoke the same language. It had forgotten how to breathe.

That city is the aging human body. The most dangerous misconception about aging is that it's the slow failure of individual parts—knees, memory, skin. This view is tidy. It gives us something to point at. "My mitochondria are damaged." "My telomeres are short." "I have too many senescent cells." But that's not the full truth. Aging, in its essence, is not the collapse of one system but the loss of harmony between systems. The decline is in communication.

This is why we must reject the idea of aging as a linear sequence of isolated failures and instead understand it as a network breakdown—a tangled mess of feedback loops where damage in one domain spills into others, bouncing unpredictably, reinforcing decay.

Let's look at just one example.

DNA damage, an unavoidable feature of life, increases with age. That much we know. But DNA damage doesn't sit quietly in the nucleus like a broken code. It sends signals. It activates a protein called p53—aptly nicknamed the "guardian of the genome"—which decides whether the cell should halt division, repair itself, or self-destruct. Often, this activation triggers cellular senescence, a state where the cell stops dividing but doesn't die. Instead, it becomes a factory for inflammatory molecules—what scientists call the Senescence-Associated Secretory Phenotype (SASP).

SASP is like an angry letter to the neighborhood. It broadcasts that the local environment is dangerous. But unlike a fire alarm, which fades when the fire is out, this signal doesn't shut off. It keeps screaming. Soon, nearby cells become agitated, too. Stem cells in the area retreat. Inflammation spreads. DNA damage increases. And the loop completes itself:

DNA damage → p53 → senescence → SASP → stem cell inhibition + inflammation → more DNA damage.

This isn't a one-way street. It's a vicious circle. A city gridlocked by its own emergency alerts.

Now take a different route through the map.

Consider the role of nutrient sensing, the biological GPS for when to grow and when to conserve. In youth, nutrient sensing is a finely tuned machine. You eat. Insulin rises. mTOR (mechanistic Target of Rapamycin) is activated, promoting protein synthesis and growth. But with age—and often, overnutrition—mTOR becomes overactive, leading to impaired autophagy, the cell's internal cleanup crew. Proteins misfold. Aggregates accumulate. Meanwhile, mitochondria—the powerhouses—begin to sputter. Their function declines. Reactive oxygen species (ROS), the toxic exhaust, increases. And guess what? ROS damages DNA.

Loop number two: Nutrient excess → mTOR activation → impaired autophagy → protein aggregates + mitochondrial dysfunction → oxidative stress → DNA damage.

In both cases, aging is not a single event but a cascade—a domino field set in a circle, not a line.

This systemslevel view radically reframes how we think about intervention. If aging were a chain, we might cut one link and halt the process. But in a web, pulling one strand merely shifts tension elsewhere. Disabling senescence might increase cancer risk. Inhibiting inflammation might slow repair. Everything is connected.

This is why early anti-aging strategies that targeted single hallmarks—say, telomere elongation alone or antioxidant floods—failed to deliver meaningful results. They were city planners fixing potholes while ignoring that the traffic system was out of sync. The traffic lights were blinking green in every direction.

Aging, at its core, is a failure of biological homeostasis—the dynamic equilibrium that maintains cellular coordination, repair timing, and tissue function. And homeostasis isn't maintained by one switch. It's upheld by orchestration. Not the perfection of each player, but the timing between them.

This is why the metaphor of a city that forgets how to breathe is so powerful. Breathing is not just the lungs. It's the brainstem's rhythm, the diaphragm's contraction, the blood's gas exchange, and even the tissuelevel sensors that say, "Enough oxygen, now exhale." You don't age because the lungs forget. You age because the signal to inhale and the capacity to exhale drift apart. A misfire here, a delay there—and the system becomes erratic.

This part of the chapter asks the reader to zoom out. Forget the damaged gene, the fatigued mitochondrion, the inflammatory molecule—for a moment. Think instead of the wiring between them. When the wires fray, when the subsystems stop sending reliable signals, the music falters. The lights flicker. The city slows.

The rest of the chapter will dive into how each of these subsystems converges. But here, we set the stage: aging is not the failure of pieces—it is the failure of synchronization. The city's buildings still stand. But its breath? That rhythm is broken.

And perhaps—just perhaps—some of it can be restored.

CONVERGENCE

In Chapter 2, we followed two siblings with dyskeratosis congenita, a rare telomeropathy. They were born with shortened telomeres, and by their late teens, they began showing signs of early aging: bone marrow failure, pulmonary fibrosis, and skin atrophy.

Their case clarified something essential—telomeres don't just cap chromosomes; they gate access to regenerative capacity. Once telomeres eroded past a threshold, stem cells in high-turnover tissues (like the lungs, gut, and skin) shut down. Cellular senescence spiked. Inflammatory markers rose. DNA damage checkpoints kicked in.

But perhaps the most instructive detail was the cascade: telomere shortening → stem cell exhaustion → tissue damage → inflammation → more DNA stress. It was not a single point of failure—it was dominoes falling across the network. Even the immune system began to mirror that of an octogenarian, years ahead of schedule. The siblings weren't just victims of a single hallmark—they illustrated how one disrupted process can entangle the others.

Later, in Chapter 7, we met the osteoarthritis patient enrolled in a senolytic trial. She could barely walk without pain. Imaging showed advanced joint degeneration. What happened next wasn't miraculous—it was instructive. After three months on a senolytic cocktail targeting p16INK4a-positive cells, her pain reduced by half. Her gait stabilized. Inflammatory markers plummeted.

Senolytics didn't just clear her senescent joint cells—they broke a loop. Fewer senescent cells meant less SASP (the senescence-associated secretory phenotype), which meant less immune overactivation and joint degradation. But the ripple effects didn't stop there: stem cells in her cartilage began to reactivate. Mitochondrial function improved in her biopsy samples. Proteostasis markers normalized slightly. The intervention had targeted one hallmark, but its effects spilled over into at least four others.

Her case highlighted a therapeutic truth: target a node in the aging network, and you may restore systemwide function—not by magic, but by alleviating pressure points that were choking feedback loops.

The chef with Parkinson's, introduced in Chapter 4, began to lose more than his culinary precision. His speech slurred. His sense of smell faded. Neurodegeneration seemed inevitable. But when researchers analyzed his cerebrospinal fluid and blood markers, they saw something unexpected—his profile showed not just α-synuclein aggregation (a proteostasis issue) but severe mitochondrial damage, oxidative stress, and a spike in inflammatory microglia.

His case illuminated one of the most elusive aspects of neurodegeneration: no single root. Misfolded proteins clogged his neural circuits, but they likely accumulated because autophagy—the cell's trash removal system—was failing. That autophagy failure was partly driven by nutrient-sensing dysregulation and chronic mTOR activation. And mitochondrial toxins weren't helping either. Even immune surveillance in the brain, usually kept in check by the bloodbrain barrier, had gone rogue, triggering further damage.

What looked like one disease—Parkinson's—was a melting pot of five hallmarks playing off each other in destructive synchrony. His treatment didn't cure the disease, but a combined approach involving caloric restriction mimetics, mitochondrial support, and anti-inflammatory nutraceuticals slowed the decline. His story became a blueprint for attacking neurodegeneration not through one lever, but several—each informed by a hallmark.

And finally, our inflammaging protagonist from Chapter 9—the 85-year-old woman overwhelmed by internal "noise." Her gut was leaking, her immune system confused.

Her blood markers read like a cytokine storm in slow motion.

Her intervention was deceptively simple: a high-fiber diet, a custom probiotic strain, and intermittent use of an IL-6 inhibitor. The results? Gut permeability reduced. Inflammatory markers dipped. Her sleep improved. Her cognition sharpened. Her T-cells began to regain some of their youthful balance.

Her case showed the profound reach of the gutimmune axis—and how treating dysbiosis could reverberate upward to recalibrate cellular senescence, mitochondrial tone, and even DNA stability through reduced oxidative stress. She didn't just feel better—her aging slowed, subtly but measurably.

Each of these lives reminds us: there is no single story of aging. The hallmarks don't act in isolation—they act in concert, sometimes harmoniously, sometimes catastrophically. understanding these intersections is the beginning of a more precise, human-centered geroscience. These are not just case studies. They are fractals of the whole picture—each one illustrating how biology's complexity can bend, and occasionally heal, through thoughtful, layered intervention.

FROM HALLMARKS TO NETWORKS

In the 1970s, a computer scientist named Gordon Bell attempted something that seemed simple but was devilishly complex: creating a personal computer network at Digital Equipment Corporation. The idea was to allow different systems to talk to each other—printers, mainframes, workstations—but the cables turned out to be the easy part. The real challenge was getting every system to speak the same language. A problem in one part of the network—say, a poorly written driver—could cause inexplicable slowdowns or crashes somewhere entirely different. Debugging it was like pulling on one thread of a spiderweb and watching the whole web twitch.

Eventually, this thinking would become the foundation of what we now call network architecture.

Aging is much the same. For decades, biologists studied individual hallmarks in isolation—one node at a time. Genomic instability. Mitochondrial dysfunction. Telomere attrition. But the deeper we dig, the more we realize that aging behaves less like a sequence and more like a circuit board. You can't just fix one transistor. Each component loops into others, forming feedback circuits that are circular, recursive, and sometimes self-destructive. When one goes wrong, it may trigger a cascade that affects the entire network.

Take the example of mTOR—a nutrientsensing pathway often described as a master regulator. Inhibiting mTOR, as with rapamycin, has been shown in animal models to enhance autophagy, a key mechanism of proteostasis. That in turn reduces protein aggregates and eases endoplasmic reticulum stress. But it doesn't stop there. Autophagy upregulation also improves mitochondrial health, reducing reactive oxygen species, which then lowers DNA damage and inflammation. That, in turn, preserves stem cell niches and may even slow telomere shortening. What looked like a single lever turns out to be a master node in a much larger regulatory network.

This shift—from hallmarks to networks—is more than just a theoretical upgrade. It's a practical revolution. In the past five years, labs have begun employing systems biology models to map these interactions in fine detail. Instead of drawing arrows on whiteboards, they're using dynamic simulations. They feed in data from transcriptomics, proteomics, and metabolomics—creating multidimensional models of cellular aging. What's emerged is a map that's messy, beautiful, and sobering.

One example is the Geroscience Systems Atlas project, which attempted to rank interventions not by their direct effect on lifespan, but by how many hallmarks they influenced across datasets. Senolytics ranked high, not because they only cleared senescent cells, but because doing so also decreased SASP-driven inflammation, reawakened dormant stem cells, and restored mitochondrial energetics. The intervention was systemic, even if its delivery was targeted.

We're now seeing the same logic play out in experimental therapeutics. Combinations like rapamycin plus NAD+ boosters. Or senolytics followed by stem cell therapy. In animal models, these cocktails often produce results that are not merely additive but synergistic—one treatment primes the system for another to take effect. It's as if the circuits can't be fixed sequentially. They have to be rebooted in parallel, with multiple switches flipped in the correct order.

This new view also challenges an old assumption: that the best anti-aging therapy is the one that targets a single dominant mechanism. That may hold true for monogenic diseases, but aging is polygenic, multilayered, and full of redundancies. Blocking one pathway may simply reroute damage through another. That's why the future lies not in monotherapies but in what systems biologists call "multi-node perturbations." You nudge several interconnected hubs gently rather than smashing one target with a sledgehammer.

There's also a new movement toward defining not just the hallmarks of aging—but the hallmarks of health.

Robust proteostasis, intact mitochondrial dynamics, balanced immune tone, flexible epigenetics. This reframing does two things. First, it gives us a positive model to aim for, not just a list of what breaks down. Second, it aligns aging science with functional restoration, not just damage control. Think of it as restoring music to a tangled radio signal, not merely patching up the wires.

In this framework, the individual hallmarks become less like standalone problems and more like symptoms of a deeper, underlying network failure—homeostatic collapse. This idea echoes what some evolutionary biologists have long argued: that aging is less a programmed decline and more a failure of repair systems to keep up with multi-source damage. The decline isn't about one node going bad. It's about the system's declining ability to keep the orchestra in tune—or, to stick to our current metaphor, to keep the grid from brownouts and surges.

Even this metaphor has its limits. The network is not mechanical—it's organic. It rewires itself. It adapts. But with age, the flexibility of that network—the biological term is "plasticity"—declines. That's why interventions like intermittent fasting, exercise, or even certain stressors (like hormesis) work so well. They don't just fix one wire. They reboot the whole mesh, often activating ancient, conserved survival circuits that tune multiple subsystems at once.

So what does this mean for the future of aging research? It means the war on aging may be less about silver bullets and more about smart orchestrations. It's not about killing senescent cells or extending telomeres or rebooting mitochondria. It's about doing all of that—but in the right order, in the right dose, in the right tissue. It's about systems control, not just symptom relief.

And perhaps most importantly, it means aging may not be a single disease to cure—but a complex process to reprogram.

ETHICAL HORIZONS AND CONTRARIAN VOICES

The year was 1997, and a sheep named Dolly had just been cloned. In science classrooms and on nightly news broadcasts, debates erupted like wildfires. Could we clone humans next? Should we? Would this mean an end to natural reproduction, a disruption of evolution, a tool for the elite? At the time, Dolly wasn't just a biological marvel—she was a moral earthquake. Everyone, from the Pope to policymakers, had something to say. Yet over the decades, the dust settled. Cloning didn't become mainstream, but it opened doors: to stem cell research, gene therapy, and now, to a far more radical question—should we treat aging?

This is not a theoretical seminar anymore. Across the globe, aging is being probed like never before. Labs from California to Kyoto are trialing interventions aimed not at diseases of old age—but at old age itself. And with this pursuit comes a new era of ethical and philosophical reflection.

At a recent roundtable in Geneva—comprising scientists, bioethicists, sociologists, and even religious scholars—a question was put to the group: If we could slow or even reverse aging, should we? The room fell silent before the debates ignited.

One camp, primarily ethicists and public health experts, argued that such breakthroughs would exacerbate inequality. Would these therapies be affordable to the global south? Would billionaires buy decades while others live handtomouth, dying at fifty? A biotech CEO responded: "That's how all medicines begin—expensive. But look at HIV treatment. Or insulin. What's rare becomes common." But the rebuttal came quickly: "Yes—but access didn't equalize by itself. It took policy, protest, and decades."

Others raised a more subtle concern—what if we succeed, but at the wrong thing? Extending life without extending vitality. Imagine an additional twenty years of frailty, cognitive decline, dependency. Aging, they argued, is not one disease but many, and prolonging it without improving quality could strain families, systems, and the very dignity of old age.

Some voiced ecological alarms. Our planet already teeters on the edge of overpopulation. What happens when millions of people live twenty years longer? Would retirement age shift to ninety? Would younger generations face job gridlock? Would resource distribution buckle? These weren't apocalyptic fears—they were logistical ones, grounded in demography, not dystopia.

But perhaps the most philosophical challenge came from a Cambridge historian: What does it mean to live a full life? If death is pushed further into the future, does urgency dissolve? Do our lives stretch meaning thin, or do we finally get enough time to pursue depth over speed?

Not everyone at the table was worried. Some, particularly the researchers developing these therapies, pointed to precedent. Every major medical breakthrough has stirred panic. Antibiotics were once seen as unnatural interference with God's will. Organ transplants were condemned as body desecration. IVF—now routine—was decried as Frankenstein science. "The moral compass of society shifts," one scientist argued. "Today's heresy is tomorrow's standard care."

Still, the contrarian scientists in the room had their doubts—not about ethics, but biology. "We haven't proven any of this in humans yet," one immunologist stated flatly. "Extending lifespan in mice is a milestone. But humans are not just bigger mice. We're more complex, more chaotic, more social."

Others raised technical concerns. Intervening in one process—like cellular senescence—might destabilize another. What if clearing senescent cells inadvertently weakens tissue repair or heightens cancer risk? "Aging is a web," said one geneticist. "Pluck one thread too hard, and the tapestry warps."

And then came the camp of the romantics—the naturalists. They argued that aging was not a defect but a feature. That the transience of life gives it meaning. "If we never feared time, would we ever value it?" one poetturned-ethicist asked. Aging, they suggested, was not a problem to be solved, but a process to be embraced.

The counterpoint? We already treat aging. Hip replacements, cataract surgery, statins—these are all interventions against the wear of time. The difference now is intent. For the first time, we're not patching symptoms. We're asking if the root processes can be modified.

The room never reached consensus. And maybe that's the point. Treating aging isn't just a scientific endeavor—it's a societal reckoning.

But here's a framing worth considering: we don't question the morality of extending health. When we vaccinate children, when we treat cancer, when we slow heart disease—we're intervening in nature. And we're doing it not to deny death, but to dignify life.

So perhaps the ethical question isn't "Should we treat aging?" but "How do we treat aging wisely, equitably, and meaningfully?" Not to create a world of immortals, but to reduce needless suffering, to compress the years of disease and decline into a narrow window, and to let people live with energy and autonomy as long as their biology allows.

The horizon of aging science is lined with promise—and shadowed by caution. But as history shows, the most transformative advances have always walked this line. The tools are coming. The question is: what kind of world will we build with them?

A NEW CHAPTER IN THE HUMAN STORY

What does it mean to rewrite the story of aging?

It doesn't mean erasing decline or freezing ourselves in time. Instead, it means compressing the most debilitating years into a narrower window—delaying their onset, softening their impact, and reclaiming more time for autonomy, creativity, and vitality.

This entire book has walked through the twelve biological hallmarks of aging: not as a grim checklist, but as threads in a woven tapestry of change. And in this final chapter, we shift the narrative entirely—from decay to dialogue.

Aging, we now understand, is a networked system in flux. It's not an inevitable slide into chaos but a conversation between repair and damage, resilience and erosion. When repair lags behind injury, dysfunction accumulates. But if we can tilt the balance—by targeting the core mechanisms like senescent cells, mitochondrial decline, deregulated nutrient sensing, and chronic inflammation—we can redirect the arc.

Recent research shows we don't need perfection. We need recalibration.

Even partial reprogramming of epigenetic markers in mice has reversed signs of aging without triggering cancer.

Senolytics have restored mobility and cognition in animals. In humans, modest interventions—like dietary change, better sleep, and pharmacologic agents such as metformin—have shown promise in extending not just life, but the quality of life. The science is early, but the signal is clear: aging is malleable.

There's a quiet radicalism in this: the idea that nature's rules might be nudged. Evolution gave us bodies optimized for reproduction and survival through middle age. That was its job. Beyond that, it handed us little guidance. And yet, humans—curious, stubborn, imaginative—have chosen to write past that evolutionary script.

In some ways, this is the most human act of all: to take what nature gave us, and then ask—what else is possible?

The hallmarks are no longer just a diagnostic framework; they are coordinates on a map we now begin to navigate. The field of geroscience—once fringe—is now foundational. The old question, "Why do we age?" is being replaced with the more urgent one: "How far can we go while preserving what makes life feel worth living?"

Science will not hand us eternal youth. That was never the promise. But it may help us avoid the worst of what we assumed was inevitable. It may offer a few extra decades of movement, of cognition, of connection—what some call "the longevity dividend."

And with that comes a responsibility: not to hoard time, but to use it wisely. To invest longer lives in richer stories.

Aging, in this new light, becomes not a solitary descent, but a communal project. Something we influence—together — *through policy, care, innovation, and ethics. The conversation moves beyond labs and into living rooms. We begin to ask: How do we redesign not just the lifespan, but the life itself?*

And so the book ends not with a conclusion, but a continuation.

A cell once frail and faltering shows signs of revival. A man once slowed by age laughs with his great-granddaughter. A researcher, notebook in hand, drafts the next experiment. A woman blows out 100 candles—eyes clear, hands steady.

If aging is the story our cells write over time, then science—cautious, brilliant, flawed—has begun to learn the grammar. And for the first time, we may be ready to edit that story—not to erase its meaning, but to lengthen its melody.

Reader-tool boxes for this chapter

CLOSER LOOK ***Geroscience — the hypothesis that defines a field***

The geroscience hypothesis, articulated most clearly by Brian Kennedy and colleagues in a 2014 Cell paper, holds that aging is the largest single risk factor for the major chronic diseases of old age — cardiovascular disease, cancer, dementia, type 2 diabetes, osteoporosis — and that addressing aging biology directly will yield broader benefits than treating each disease in isolation.

The argument is partly biological and partly mathematical. If you cured all cardiovascular disease tomorrow, average life expectancy in the United States would rise by perhaps four years. If you cured all cancer, similarly. But the same patient who is saved from a heart attack will likely die of dementia, and the patient saved from cancer will likely die of cardiovascular disease. The geroscience hypothesis bets that a treatment which slows aging modestly — even by 7% — would compress the entire risk distribution simultaneously, yielding healthspan gains that single-disease cures cannot match.

Kennedy et al. (2014). Geroscience: Linking aging to chronic disease. Cell, 159(4), 709–713.

CLOSER LOOK ***Compression of morbidity — the goal nobody talks about enough***

In 1980, Stanford rheumatologist James Fries published a paper in the New England Journal of Medicine arguing that the goal of aging research should not be to extend the maximum lifespan but to compress the period of disability and disease at the end of life into a smaller window. If the average age of first chronic disease could be pushed later faster than life expectancy itself rose, the years of decline would shrink — even if the years of life held steady.

Forty-five years on, this remains the most realistic operational definition of healthy aging. It is also the metric most longevity interventions are quietly designed against. The dream is not to live to 130. It is to be functional at 90, sharp at 85, working at 75, and to enter the last chapter not as a slow decline of fifteen years but as a brief one of two. Compression, not extension.

Fries (1980). Aging, natural death, and the compression of morbidity. New England Journal of Medicine, 303(3), 130–135.

EASY ACCESS ***Why the skeptics matter***

Every emerging field accumulates a layer of overstatement. Aging is no exception. Marketing has run ahead of mechanism in supplements, in private clinics, in age-reversal startups with names that promise more than the science can deliver. The skeptics — biostatisticians who ask whether the lifespan extensions in mice translate, ethicists who ask who pays and who benefits, demographers who note that long-lived populations are not always healthy ones — are not the enemies of longer healthspans. They are the quality control. A field that listens to its critics matures faster than one that doesn't.

CLOSER LOOK ***What the data actually support, today***

The intervention with the strongest evidence in humans, by a wide margin, is regular physical activity. Resistance training preserves muscle, bone, and stem cell function. Aerobic training preserves cardiovascular and mitochondrial capacity. Combined training, started at any age, attenuates almost every measurable hallmark — including markers of inflammation, telomere stability, and epigenetic age.

Beyond exercise, the human evidence narrows quickly. Caloric restriction has solid mechanistic support and modest human data. Time-restricted eating has growing but not definitive support. Rapamycin, metformin, NAD+ precursors, senolytics — all are interesting; none are yet ready as broad recommendations for healthy adults. The honest summary is that the science is moving fast, the marketing is moving faster, and the gap between them is where careful readers, like you, can do the most good.

Garatachea et al. (2015). Exercise attenuates the major hallmarks of aging. Rejuvenation Research, 18(1), 57–89.

KEY TERMS *Vocabulary of the closing chapter*

Geroscience hypothesis — The claim that targeting aging biology yields broader health gains than targeting any single age-related disease.

Compression of morbidity — Pushing the onset of chronic disease closer to the age of death, narrowing the disabled years.

Healthspan — Years lived in good function. Distinct from lifespan.

Longevity dividend — The economic and social returns expected from interventions that slow aging in populations.

Hallmark network — The view that the twelve hallmarks of aging interact, reinforce each other, and are best targeted in combination.

REFERENCES

Abegglen, L. M., Caulin, A. F., Chan, A., Lee, K., Robinson, R., Campbell, M. S., Kiso, W. K., Schmitt, D. L., Waddell, P. J., Bhaskara, S., Jensen, S. T., Maley, C. C., & Schiffman, J. D. (2015). Potential mechanisms for cancer resistance in elephants and comparative cellular response to DNA damage in humans. JAMA, 314(17), 1850–1860.

https://doi.org/10.1001/jama.2015.13134

Andreux, P. A., Blanco-Bose, W., Ryu, D., Burdet, F., Ibberson, M., Aebischer, P., Auwerx, J., Singh, A., & Rinsch, C. (2019). The mitophagy activator urolithin A is safe and induces a molecular signature of improved mitochondrial and cellular health in humans. Nature Metabolism, 1(6), 595–603. https://doi.org/10.1038/s42255-019-0073-4

Baker, D. J., Wijshake, T., Tchkonia, T., LeBrasseur, N. K., Childs, B. G., van de Sluis, B., Kirkland, J. L., & van Deursen, J. M. (2011). Clearance of p16Ink4a-positive senescent cells delays ageing-associated disorders. Nature, 479(7372), 232–236. https://doi.org/10.1038/nature10600

Bertuch, A. A. (2016). The molecular genetics of the telomere biology disorders. RNA Biology, 13(8), 696–706. https://doi.org/10.1080/15476286.2015.1094596

Bjelakovic, G., Nikolova, D., Gluud, L. L., Simonetti, R. G., & Gluud, C. (2007). Mortality in randomized trials of antioxidant supplements for primary and secondary prevention: Systematic review and meta-analysis. JAMA, 297(8), 842–857. https://doi.org/10.1001/jama.297.8.842

Brack, A. S., & Rando, T. A. (2008). Aging of the skeletal muscle stem cell niche. Methods in Enzymology, 449, 33–61. https://doi.org/10.1016/S0076-6879(08)02402-7

Conboy, I. M., Conboy, M. J., Wagers, A. J., Girma, E. R., Weissman, I. L., & Rando, T. A. (2005). Rejuvenation of aged progenitor cells by exposure to a young systemic environment. Nature, 433(7027), 760–764.

https://doi.org/10.1038/nature03260

Coppé, J.-P., Patil, C. K., Rodier, F., Sun, Y., Muñoz, D. P., Goldstein, J., Nelson, P. S., Desprez, P.-Y., & Campisi, J. (2008). Senescence-associated secretory phenotypes reveal cell-nonautonomous functions of oncogenic RAS and the p53 tumor suppressor. PLoS Biology, 6(12), e301. https://doi.org/10.1371/journal.pbio.0060301

Eisenberg, T., Knauer, H., Schauer, A., Büttner, S., Ruckenstuhl, C., Carmona-Gutierrez, D., Ring, J., Schroeder, S., Magnes, C., Antonacci, L., Fussi, H., Deszcz, L., Hartl, R., Schraml, E., Criollo, A., Megalou, E., Weiskopf, D., Laun, P., Heeren, G., ... Madeo, F. (2009). Induction of autophagy by spermidine promotes longevity. Nature Cell Biology, 11(11), 1305–1314. https://doi.org/10.1038/ncb1975

Fries, J. F. (1980). Aging, natural death, and the compression of morbidity. New England Journal of Medicine, 303(3), 130–135.

https://doi.org/10.1056/NEJM198007173030304

Furman, D., Campisi, J., Verdin, E., Carrera-Bastos, P., Targ, S., Franceschi, C., Ferrucci, L., Gilroy, D. W., Fasano, A., Miller, G. W., Miller, A. H., Mantovani, A., Weyand, C. M., Barzilai, N., Goronzy, J. J., Rando, T. A., Effros, R. B., Lucia, A., Kleinstreuer, N., & Slavich, G. M. (2019). Chronic inflammation in the etiology of disease across the life span. Nature Medicine, 25(12), 1822–1832.

https://doi.org/10.1038/s41591-019-0675-0

Garatachea, N., Pareja-Galeano, H., Sanchis-Gomar, F., Santos-Lozano, A., Fiuza-Luces, C., Morán, M., Emanuele, E., Joyner, M. J., & Lucia, A. (2015). Exercise attenuates the major hallmarks of aging. Rejuvenation Research, 18(1), 57–89. https://doi.org/10.1089/rej.2014.1623

Gomez-Cabrera, M.-C., Domenech, E., Romagnoli, M., Arduini, A., Borras, C., Pallardo, F. V., Sastre, J., & Viña, J. (2008). Oral administration of vitamin C decreases muscle mitochondrial biogenesis and hampers training-induced adaptations in endurance performance. American Journal of Clinical Nutrition, 87(1), 142–149.

https://doi.org/10.1093/ajcn/87.1.142

Harman, D. (1956). Aging: A theory based on free radical and radiation chemistry. Journal of Gerontology, 11(3), 298–300. https://doi.org/10.1093/geronj/11.3.298

Harrison, D. E., Strong, R., Sharp, Z. D., Nelson, J. F., Astle, C. M., Flurkey, K., Nadon, N. L., Wilkinson, J. E., Frenkel, K., Carter, C. S., Pahor, M., Javors, M. A., Fernandez, E., & Miller, R. A. (2009). Rapamycin fed late in life extends lifespan in genetically heterogeneous mice. Nature, 460(7253), 392–395. https://doi.org/10.1038/nature08221

Hartl, F. U., Bracher, A., & Hayer-Hartl, M. (2011). Molecular chaperones in protein folding and proteostasis. Nature, 475(7356), 324–332.

https://doi.org/10.1038/nature10317

Hayflick, L., & Moorhead, P. S. (1961). The serial cultivation of human diploid cell strains. Experimental Cell Research, 25(3), 585–621. https://doi.org/10.1016/0014-4827(61)90192-6

Hickson, L. J., Langhi Prata, L. G. P., Bobart, S. A., Evans, T. K., Giorgadze, N., Hashmi, S. K., Herrmann, S. M., Jensen, M. D., Jia, Q., Jordan, K. L., Kellogg, T. A., Khosla, S., Koerber, D. M., Lagnado, A. B., Lawson, D. K., LeBrasseur, N. K., Lerman, L. O., McDonald, K. M., McKenzie, T. J., ... Kirkland, J. L. (2019). Senolytics decrease senescent cells in humans: Preliminary report from a clinical trial of Dasatinib plus Quercetin in individuals with diabetic kidney disease. EBioMedicine, 47, 446–456.

https://doi.org/10.1016/j.ebiom.2019.08.069

Horvath, S. (2013). DNA methylation age of human tissues and cell types. Genome Biology, 14(10), R115.

https://doi.org/10.1186/gb-2013-14-10-r115

Jaskelioff, M., Muller, F. L., Paik, J.-H., Thomas, E., Jiang, S., Adams, A. C., Sahin, E., Kost-Alimova, M., Protopopov, A., Cadiñanos, J., Horner, J. W., Maratos-Flier, E., & DePinho, R. A. (2011). Telomerase reactivation reverses tissue degeneration in aged telomerase-deficient mice. Nature, 469(7328), 102–106. https://doi.org/10.1038/nature09603

Kennedy, B. K., Berger, S. L., Brunet, A., Campisi, J., Cuervo, A. M., Epel, E. S., Franceschi, C., Lithgow, G. J., Morimoto, R. I., Pessin, J. E., Rando, T. A., Richardson, A., Schadt, E. E., Wyss-Coray, T., & Sierra, F. (2014). Geroscience: Linking aging to chronic disease. Cell, 159(4), 709–713. https://doi.org/10.1016/j.cell.2014.10.039

Kenyon, C., Chang, J., Gensch, E., Rudner, A., & Tabtiang, R. (1993). A C. elegans mutant that lives twice as long as wild type. Nature, 366(6454), 461–464.

https://doi.org/10.1038/366461a0

Lu, Y., Brommer, B., Tian, X., Krishnan, A., Meer, M., Wang, C., Vera, D. L., Zeng, Q., Yu, D., Bonkowski, M. S., Yang, J.-H., Zhou, S., Hoffmann, E. M., Karg, M. M., Schultz, M. B., Kane, A. E., Davidsohn, N., Korobkina, E., Chwalek, K., ... Sinclair, D. A. (2020). Reprogramming to recover youthful epigenetic information and restore vision. Nature, 588(7836), 124–129. https://doi.org/10.1038/s41586-020-2975-4

López-Otín, C., Blasco, M. A., Partridge, L., Serrano, M., & Kroemer, G. (2023). Hallmarks of aging: An expanding universe. Cell, 186(2), 243–278.

https://doi.org/10.1016/j.cell.2022.11.001

Mattison, J. A., Colman, R. J., Beasley, T. M., Allison, D. B., Kemnitz, J. W., Roth, G. S., Ingram, D. K., Weindruch, R., de Cabo, R., & Anderson, R. M. (2017). Caloric restriction improves health and survival of rhesus monkeys. Nature Communications, 8, 14063.

https://doi.org/10.1038/ncomms14063

Niedernhofer, L. J., Gurkar, A. U., Wang, Y., Vijg, J., Hoeijmakers, J. H. J., & Robbins, P. D. (2018). Nuclear genomic instability and aging. Annual Review of Biochemistry, 87, 295–322. https://doi.org/10.1146/annurev-biochem-062917-012239

Ocampo, A., Reddy, P., Martinez-Redondo, P., Platero-Luengo, A., Hatanaka, F., Hishida, T., Li, M., Lam, D., Kurita,

M., Beyret, E., Araoka, T., Vazquez-Ferrer, E., Donoso, D., Roman, J. L., Xu, J., Rodriguez Esteban, C., Nuñez, G., Nuñez Delicado, E., Campistol, J. M., ... Belmonte, J. C. I. (2016). In vivo amelioration of age-associated hallmarks by partial reprogramming. Cell, 167(7), 1719–1733.e12.

https://doi.org/10.1016/j.cell.2016.11.052

Pickrell, A. M., & Youle, R. J. (2015). The roles of PINK1, parkin, and mitochondrial fidelity in Parkinson's disease. Neuron, 85(2), 257–273.

https://doi.org/10.1016/j.neuron.2014.12.007

Ridker, P. M., Everett, B. M., Thuren, T., MacFadyen, J. G., Chang, W. H., Ballantyne, C., Fonseca, F., Nicolau, J., Koenig, W., Anker, S. D., Kastelein, J. J. P., Cornel, J. H., Pais, P., Pella, D., Genest, J., Cifkova, R., Lorenzatti, A., Forster, T., Kobalava, Z., ... Glynn, R. J. (2017). Antiinflammatory therapy with canakinumab for atherosclerotic disease. New England Journal of Medicine, 377(12), 1119–1131.

https://doi.org/10.1056/NEJMoa1707914

Ristow, M., Zarse, K., Oberbach, A., Klöting, N., Birringer, M., Kiehntopf, M., Stumvoll, M., Kahn, C. R., & Blüher, M. (2009). Antioxidants prevent health-promoting effects of physical exercise in humans. Proceedings of the National Academy of Sciences, 106(21), 8665–8670.

https://doi.org/10.1073/pnas.0903485106

Sahin, E., & DePinho, R. A. (2010). Linking functional decline of telomeres, mitochondria and stem cells during ageing. Nature, 464(7288), 520–528.

https://doi.org/10.1038/nature08982

Selkoe, D. J., & Hardy, J. (2016). The amyloid hypothesis of Alzheimer's disease at 25 years. EMBO Molecular Medicine, 8(6), 595–608. https://doi.org/10.15252/emmm.201606210

Sender, R., Fuchs, S., & Milo, R. (2016). Revised estimates for the number of human and bacteria cells in the body. PLoS Biology, 14(8), e1002533.

https://doi.org/10.1371/journal.pbio.1002533

Tyner, S. D., Venkatachalam, S., Choi, J., Jones, S., Ghebranious, N., Igelmann, H., Lu, X., Soron, G., Cooper, B., Brayton, C., Park, S. H., Thompson, T., Karsenty, G., Bradley, A., & Donehower, L. A. (2002). p53 mutant mice that display early ageing-associated phenotypes. Nature, 415(6867), 45–53. https://doi.org/10.1038/415045a

Vousden, K. H., & Lane, D. P. (2007). p53 in health and disease. Nature Reviews Molecular Cell Biology, 8(4), 275–283. https://doi.org/10.1038/nrm2147

Yang, J.-H., Hayano, M., Griffin, P. T., Amorim, J. A., Bonkowski, M. S., Apostolides, J. K., Salfati, E. L., Blanchette, M., Munding, E. M., Bhakta, M., Chew, Y. C., Guo, W., Yang, X., Maybury-Lewis, S., Tian, X., Ross, J. M., Coppotelli, G., Meer, M. V., Rogers-Hammond, R., … Sinclair, D. A. (2023). Loss of epigenetic information as a cause of mammalian aging. Cell, 186(2), 305–326.

https://doi.org/10.1016/j.cell.2022.12.027

Zhang, G., Li, J., Purkayastha, S., Tang, Y., Zhang, H., Yin, Y., Li, B., Liu, G., & Cai, D. (2013). Hypothalamic programming of systemic ageing involving IKK-β, NF-κB and GnRH. Nature, 497(7448), 211–216.

https://doi.org/10.1038/nature12143

APPENDIX A

OTHER BOOKS IN THE F(X) LONGEVITY SERIES

Decode Aging. Reclaim Function.

FOUNDATION SERIES — THE BIOLOGY OF AGING

Unlock

Systems Biology, Aging Mechanisms, and Functional Decline

Aging is a coordinated systems breakdown that can be measured and influenced.

Cool Down

Inflammation, Immune Signaling, and Chronic Disease Acceleration

Chronic inflammation drives the majority of age-related diseases.

Power Up

Mitochondrial Bioenergetics, Cellular Energy, and Longevity Output

Energy decline is the earliest measurable signal of aging.

Reboot

Glymphatic Clearance, Brain Detoxification, and Cognitive Preservation

Brain waste clearance during sleep determines long-term cognitive resilience.

Stay Strong

Muscle Physiology, Protein Turnover, and Functional Capacity

Muscle loss is a primary driver of frailty and mortality.

Stronger Still

Bone Remodeling, Mineral Density, and Structural Integrity

Bone decline reflects systemic metabolic and hormonal dysfunction.

Defend

Immune Surveillance, Host Defense, and Aging Vulnerability

Aging shifts immunity toward chronic activation and reduced precision.

Reset

Cellular Senescence, SASP Signaling, and Tissue Dysfunction

Senescent cells propagate aging through inflammatory signaling.

CONTROL SYSTEMS — METABOLIC AND SIGNALING NETWORKS

Fire Up

Metabolic Rate, Thermogenesis, and Energy Throughput

Metabolic slowdown reflects modifiable changes in energy flux.

Signal

Cellular Communication, Signal Transduction, and System Coordination

Aging is a failure of cellular communication.

Rewire

Neuroplasticity, Dopaminergic Signaling, and Behavioral Drive

Motivation and cognition are biologically regulated systems.

The Metabolic Switch

TOR, AMPK, Autophagy, and Nutrient-Sensing Networks

Longevity depends on repair versus accumulation signaling.

The Metabolic Signal

Hormonal Regulation, Metabolic Control, and Longevity Pathways

Metabolic dysfunction underlies most chronic disease.

INTERVENTIONS — TARGETED OPTIMIZATION

The Peptide Protocol

Regenerative Signaling, Peptide Therapeutics, and Tissue Repair

Peptides enable targeted restoration of cellular function.

The Testosterone Reset

Androgen Signaling, Anabolic Balance, and Performance Longevity

Testosterone regulates far more than reproduction.

The Estrogen Code

Estrogen Signaling, Mitochondrial Protection, and Female Longevity

Estrogen is a central regulator of systemic health.

ADAPTATION — STRESS AND PERFORMANCE PHYSIOLOGY

The Adaptation Code

Stress Physiology, Hormetic Load, and Biological Resilience

Adaptation depends on dose, timing, and recovery.

The Recovery Principle

Sleep Architecture, Repair Systems, and Regenerative Capacity

Recovery determines whether stress builds or breaks the system.

The Circadian Code

Biological Rhythms, Clock Genes, and Temporal Optimization

Timing is a primary regulator of physiology.

The Oxygen System

Respiration, CO_2 Physiology, and Cellular Efficiency

Gas exchange governs mitochondrial performance.

PHYSIOLOGIC BOTTLENECKS — SYSTEM INTEGRATION POINTS

Clear Flow

Renal Filtration, Fluid Dynamics, and Metabolic Clearance

Kidney function regulates systemic aging through filtration and pressure.

The Vascular Code

Endothelial Function, Blood Flow, and Tissue Perfusion

All systems depend on vascular integrity.

The Gut-Immune Axis

Microbiome Ecology, Barrier Function, and Inflammatory Control

The gut defines systemic inflammation and immune tone.

The Detox Network

Hepatic Biotransformation, Toxic Load, and Systemic Balance

Detoxification is a continuous biochemical necessity.

INTEGRATION

Human Code

Systems Integration, Network Physiology, and Lifespan Optimization

Longevity emerges from coordinated system-wide function.

www.ingramcontent.com/pod-product-compliance
Ingram Content Group UK Ltd.
Pitfield, Milton Keynes, MK11 3LW, UK
UKHW020126290926
15471UKWH00003B/315

9 798295 701887